Lectures on Torah and Modern Physics

The Lectures in Kabbalah Series

Lectures on Torah and Modern Physics

by

Harav Yitzchak Ginsburgh

Transcribed, edited, and annotated
by **Rabbi Moshe Genuth**

Gal Einai

THE LECTURES IN KABBALAH SERIES

LECTURES ON TORAH AND MODERN PHYSICS

Harav Yitzchak Ginsburgh

Edited by Rabbi Moshe Genuth

Printed in the United States of America and Israel
First Edition

For information:

Israel: GAL EINAI
PO Box 1015
Kfar Chabad 60840
tel. (in Israel): 1-700-700-966
tel. (from abroad): 972-3-9608008
email: books@inner.org
Web: www.inner.org
Blog: www.rabbiginsburgh.com

GAL EINAI produces and publishes books, pamphlets, audiocassettes and videocassettes by Rabbi Yitzchak Ginsburgh. To receive a catalog of our products in English and/or Hebrew, please contact us at any of the above addresses, email orders@inner.org or call our orders department in Israel.

Layout and Cover design: David Hillel
Cover image: Particle collision event recorded at the LHC,
courtesy of CERN (www.cern.ch)

ISBN: 978-965-7146-72-9

"נכון שיכתוב בצורת ספר השיעורים שלומד.
בברכה להצלחה"

"...It would be proper to publish your classes in book form.
With blessings for success..."

— *from a letter from the Lubavitcher Rebbe*
to the author, Elul 5741

Table of Contents

Preface

The lectures appearing in this volume were given by Rabbi Yitzchak Ginsburgh in Los Angeles, at the then Ahavat Shalom Synagogue, over the course of 3 days, March 18-20, 2007. The 3 day seminar was titled "Beyond the Physical Universe" and it focused on the relationship between modern physics and Torah.

Following the seminar, transcripts were made from recordings of the lectures and both video recordings and the transcripts were posted on Gal Einai Institute's website (www.inner.org).

The present volume is a lightly edited version of those transcripts. Rabbi Ginsburgh did not edit these transcripts and they did not go through the rigorous and in-depth process that our publications usually do. We believe that even prior to an in-depth review and expansion, the material Rabbi Ginsburgh presented at this seminar is invaluable and should be made available to readers interested in the relationship between Torah and science.

Where relevant, the original Hebrew of verses or idioms is noted. When a Hebrew word appears in bold-print, it indicates that the numerical value (*gematria*) of that word is being considered.

In order to avoid confusion, God's Names in Hebrew appear using the proper spelling, with a dash separating the letters, for halachic reasons. We follow the halachic convention of writing out God's Name in English without a dash. Some, but not all of the sources noted in the lectures have been cited in endnotes, as well as some explanatory points on the discussion.

We have made an effort to retain Rabbi Ginsburgh's speaking style in this volume and very little has been changed.

♦ ♦ ♦

Rabbi Ginsburgh has a unique place among the Torah scholars and leaders of our generation, part of it due to his ability to clearly see where Torah and science do indeed unite. Perhaps even more incredible is his ability to recognize those points in which the Torah can guide science in mankind's quest to understand the secrets of the physical universe.

We pray the Almighty provide us with the ability to continue publishing Rabbi Ginsburgh's voluminous works on Torah and science, covering practically all fields of scientific inquiry.

♦ ♦ ♦

During 2007-8, Rabbi Ginsburgh was able to visit Los Angeles every 3 months, thanks to the founding of the Los Angeles Inner Torah Institute by his students: Rabbi Shaya Eichenblatt, Prof. Eliezer Zeiger, Mrs. Chana Rachel Schusterman, and Mrs. Olivia Schwartz, all well-known and active members of the Chabad-Lubavitch community in Los Angeles.

The Inner Torah Institute and hence Rabbi Ginsburgh's regular visits to Los Angeles that year, were made possible by a generous donation from Mr. David Kaplan. We would like to take the opportunity to thank everyone who put in a great deal of hard work to make these visits possible. They all have a part in bringing the Torah's inner light to the edge of Western civilization.

Special thank to Mrs. Rachel Gordon for proofreading the transcripts.

May the Almighty indeed open our eyes to see the wonders of His Torah, allowing us to see beyond the physical dimension of our universe.

Moshe Genuth

Chanukah 5773

Lecture 1

In this series of lectures we will connect the issues of modern physics with the inner dimension of the Torah. One of the benefits to be gained from this unification is that the concepts and ideas that form the foundation of modern physics, which appear at first glance to be far removed from our everyday lives, are actually most applicable.

Modern physics is counter-intuitive

The first topic we will address is counter-intuition. It is a simple and well-known fact that all of the most important and successful theories presented and used in modern physics are counter-intuitive. The fact that the accepted scientific narrative of the past century has been utilizing counter-intuitive ideas is a very significant one, indicating that humanity in general is approaching a level that is above and beyond the style of human logic that has been applied since time immemorial.[1]

On a different front, the realization of our long anticipation of the complete and true redemption by Mashiach is imminent. Like science, the redemption that awaits us is counter-intuitive to the redemption that we have been expecting, so much so that the Alter Rebbe of Chabad was known to say that the Mashiach that everyone is waiting for will never come, and the Mashiach that will come, nobody is waiting for! So now, let us begin our journey into counter-intuition and try to see how counter-intuition applies in our lives.

Counter-intuition in Kabbalah

Kabbalah supplies us with the tools for understanding counter-intuition in a much more sophisticated way than does science. By counter-intuition, science means anything that runs against the grain of our common-sense. Special relativity, general relativity, quantum mechanics, and string theory (to a degree), are all theories that run against our common-sense and are therefore considered by science to be counter-intuitive.

The first insight that Kabbalah offers is that everything is relative. This is similar to the conclusion of the first major theory of modern science, relativity. Of course, if "everything is relative" then counter-intuition is relative as well. This means that the blanket statement that counter-intuition is anything that goes against common-sense is not accurate, because common-sense itself is relative!

In the language of Kabbalah, every mind-space, or level of consciousness, is called a World of the and every World has its own "common-sense"[2] [we capitalize "World" in the Kabbalistic sense, in order to differentiate it from the word's more common meaning; ed.] Thus, given the particular mind-space that I currently inhabit, I would consider the common-sense found at the level above mine to be counter-intuitive to my own. As we progress through each level of consciousness, from Kabbalistic World to Kabbalistic World, our common-sense undergoes a transformation and what is perceived as common-sense in one World becomes absolutely counter-intuitive to common-sense in another World. When we progress from World to World, which are actually levels of consciousness everything that we held to be true previously transforms into something new.

Those readers familiar with our method of teaching know that as we survey a new subject, one of our goals is to describe the appropriate Kabbalistic models that correspond with the particular topic at hand. If the model indeed fits and corresponds correctly, then not only do we gain insight into the model, but we immediately improve our understanding of the topic under consideration.

The model that we have chosen to use to enhance our understanding of counter-intuition is indeed based on the model of the Kabbalistic Worlds, specifically the three lower Worlds which are known as Action, Formation, and Creation. Or, in Hebrew: *Asiyah*, *Yetzirah*, and *Beri'ah*.

The basic state of human consciousness is in the World of Action or more specifically, in the physical dimension of the World of Action. Above our World is the World of Formation and above that is the World of Creation. The common denominator of these three lower Worlds is that in them everything has some degree of self-consciousness, causing it to feel separate from the Almighty; the absolute and unique One.

Above these three Worlds is the World of Emanation (*Atzilut*) in which there is no distinct self-consciousness at all; everything is purely Divine and there is no separation (or feeling of separation) possible from the Creator.

As mentioned above, every World has its own version of what is intuitive and what common-sense is and anything that contradicts it is considered "counter-intuitive." When an apparently counter-intuitive insight is attained in any of the three lower Worlds, first we should ascertain that it is indeed counter-intuitive relative to that World's common-sense.

The picture that we should have in mind as we consider how new insights are attained is that light (i.e., energy, in this case the energy that makes up the counter-intuitive thought or idea) from the World of Emanation has descended and permeated a lower World. Each of the physical theories of the 20th century that has been successfully incorporated into modern science is like a ray of light that has filtered into human consciousness from a higher level of reality. Yet, even though it seems that these theories run against the grain of our common-sense, if you ask a student of Kabbalah and Chassidut, one who is truly versed in the wisdom of the World of Emanation, whether these "counter-intuitive" theories make sense or not, he will tell you that they are in perfect harmony with what is described in the holy books about the reality of the World of Emanation. We hope that this point will be made very clear through the scientific examples we will now consider.

"The opposite makes sense"

The message that scientific progress has delivered in the last century is that the more our understanding of nature increases, the more our common-sense is challenged. In other words, scientific progress means challenging our common-sense about physical events in the world.

If this message about changing our common-sense is indeed true, we would expect to find it echoed in a verse from the Bible or in the literature of the sages. Indeed, there is an idiom the Talmudic sages use to express exactly this idea. In the original Aramaic it is pronounced, "*ipcha mistabra*" (אִיפְּכָא מִסְתַּבְּרָא). This phrase literally means: "the opposite makes sense," or in other words, the truth—what truly makes sense—is exactly the opposite of what you initially thought makes sense. This idiom appears exactly 19 times in the Babylonian Talmud, and we will have more to say about this phenomenon later.

The initial letters of this idiom spell the Hebrew word for "mother" (אֵם), hinting at its connection with what in Kabbalah is known as the mother figure, or *partzuf Ima*. The mother figure is associated with the *sefirah* of understanding, which is related to common-sense and the analytical skills it employs. The initials of this idiom thus echo the concept of common-sense, that which it is in fact challenging.

Having explained the mother principle, let us add something that will be useful later on. The father figure in Kabbalah is a connotation for the *sefirah* of wisdom, which represents direct intuition, a state of intellectual insight in which it is impossible to experience that the opposite might be true. Thus, the father principle in Kabbalah represents an insight that can never be refuted.[3]

What is the function of the Talmudic idiom, "*ipcha mistabra*?" In a Talmudic debate one sage will suggest that, "Given that x is true, such and such follows." Another sage may then argue "*ipcha mistabra*"—"Given that x is true, the opposite is the case." *Ipcha mistabra* is thus a rhetorical tool that challenges not only conclusions, but the very reasoning being used to arrive at those conclusions. In a certain sense, it is not the facts that are being challenged but their correct interpretation. This therefore exemplifies the idea stated above that the manner in which I am used to perceiving reality defines my "common-sense," but someone in a different mind-space will find my interpretation a challenge to his perception of reality and hence to his notion of what common-sense is.

The sages compiled two great anthologies of their knowledge called the Babylonian Talmud and the Jerusalem Talmud (the latter is also known as the Talmud of the land of Israel[4]). Many of the same sages appear in both editions of the Talmud, and of course the discourse in both focuses on the same basic text, the Mishnah. Yet, despite the many parallels between the two Talmuds, interestingly, the idiom *ipcha mistabra* appears only in the Babylonian, but not once in the Jerusalem Talmud. What is the reason for this?

The difference between these two works is that the discourse in the Jerusalem Talmud is concise and promptly arrives at its conclusions, whereas the Babylonian Talmud's discourse is drawn-out and argumentative in its style. Many arguments and counter-arguments are presented in the Babylonian Talmud before a final conclusion is reached, which is why it is so much longer than the Jerusalem Talmud.

Both the Jerusalem Talmud and the Babylonian illustrate the refined intellect, but the differences between them, such as this one, indicate that a refined intelligence has two different aspects. In Kabbalistic terminology, the two aspects of the intellect (there is also a third, which connects the intellect with the emotions) are identified with the two *sefirot* called wisdom and understanding and corresponding with the father and mother principles, respectively, as mentioned earlier.

The Jerusalem Talmud echoes the insightful intelligence of the father principle, the *sefirah* of wisdom. In the model of four Worlds, wisdom corresponds to the World of Emanation, where God's Presence is absolutely clear and therefore there is no experience of being separate from God. Because of the constant experience of God's all-encompassing omnipresence inherent in the mind-set of Emanation, the *sefirah* of wisdom is clear, concise, and most importantly for our purposes here, intuitively secure. The discourse in the Jerusalem Talmud leaves no room for challenges to its conclusions, because they are not arrived at by a critical methodology—they simply are!

But the Babylonian Talmud echoes the mother principle, the *sefirah* of understanding. As we touched upon earlier, this *sefirah* reflects the mind-set of the World of Creation (one level below the World of Emanation), where consciousness does include the relative experiences of selfhood. The more the person inhabiting this mind-set is aware of his own being, the more open he is to seeing how relative his methodology and conclusions seem to the subjective experiences of others.

The mind-set of Creation includes relative statements about right and wrong. Therefore, it leaves room for actually being wrong, even in intuitive matters—those things that seem to just be plain common-sense. What follows is that if you inhabit the mind-set of Creation, your first intuition might be correct, or it might be incorrect; in fact, it is often incorrect.

This does not imply that the sages of the Babylonian Talmud were at a lower level than the sages of the Jerusalem Talmud and that therefore their intuition can be challenged. Anyone designated a sage in either Talmud lived in the mind-space of Emanation. Like we said before, many of the sages are mentioned in both editions of the Talmud. It is rather that because the Babylonian Talmud was written outside the land of Israel, its editorial style reflects the mind-set of the World of Creation. Therefore, in the context of the dialogue recorded in the Babylonian Talmud, even the intuition of a great sage may be challenged. It may even be found, in the course of a Talmudic argument, to be incorrect; even completely opposite from the truth, "*ipcha mistabra*."

The feminine and counter-intuition

Let's take another step in looking at *ipcha mistabra*, the idiom that represents counter-intuition in Judaism. If we search the entire Babylonian Talmud, we find that this idiom is used exactly 19 times; the numerical value of the name "Eve" (חַוָּה), in Hebrew.[5] Eve was the first woman and is referred to in the Torah as the "mother of all life." This is especially significant when we recall that the initials of *ipcha mistabra* spell "mother," in Hebrew.

What this reveals is an implicit but essential relationship between counter-intuition and the Torah's first female figure, who more than any other character in the Bible, represents the mother principle. The first thing we learn from this connection is that indeed the woman is the *ipcha mistabra* of the man—female understanding posits a counter-intuitive alternative to male common-sense.[6]

But, more deeply, we often refer to *feminine intuition*, which is sometimes even hailed as being more potent than its masculine counterpart. Yet, unlike the adamant male intuition, a woman's intuition is capable of suddenly and completely reversing itself, not only as a response to changing circumstances, but more importantly because the woman who originally expressed the intuition suddenly feels that the exact opposite is true. In a sense, the feminine mind is tuned into intuition (the faculty that is the father principle) and adopts an intuitive stance towards life. But, this is also the reason why feminine intuition knows that it is not always correct. The female mind is thus both open to trusting its intuition, but just as open to completely reversing it by 180 degrees.

Counter-intuition and *teshuvah*

Those familiar with our methodology know that there is great deal of insight to be gained from looking at the numerical value, called the *gematria*,[7] of Hebrew words.[8] The *gematria* of the idiom *ipcha mistabra* (אִיפְּכָא מִסְתַּבְּרָא) is 815. This is also the *gematria* of the predicate *ba'al teshuvah* (בַּעַל תְּשׁוּבָה), which literally means: "a master of repentance." A *ba'al teshuvah* is someone who has returned to the Almighty through the Torah, someone who has decided to master the art of self-transformation in order to manifest a bond with God. Self-transformation requires a switch in one's consciousness. A *ba'al teshuvah* has to embrace ideas and actions that are counter-intuitive. To seek God from his or her present context, the *ba'al teshuvah* has to rise to a new level of understanding that is counter-intuitive to his or her current, common-sense approach to life.

Teshuvah—return to God—applies differently to individuals with different levels of religious observance. The more religious a person is, the more his *teshuvah* can be charted along a straight line, measuring increased attention to the details of observance and the time spent studying Torah. For the observant Jew, *teshuvah* may not involve the same type of counter-intuitive leap that it does for the Jewish returnee, the *ba'al teshuvah* of our times. Therefore, most observant Jews will point to the month of *Elul*, known as the month of *teshuvah*, as an auspicious time for the observant individual to "do" *teshuvah*.

But, a cornerstone of Chassidic teachings on spiritual consciousness is that regardless of an individual's level of observance or commitment to Torah as a way of life, he should always seek to become a full-fledged *ba'al teshuvah*; meaning that he should always see himself as still distant from God. In the words of one Chassidic master,[9] the more a person feels a sense of spiritual accomplishment and nearness to God, the more spiritually distant he actually is. This foundation of Chassidic guidance is based on the philosophical adage that a person should be practicing *teshuvah* every single day of his life.[10] Chassidic masters interpret this to mean that relative to God's absolute nature, we are all equidistant from Him. Therefore, we are all equally in need of a counter-intuitive switch in our thinking in order to better our actions. The path to spiritual progress is not traversed by a series of successive, forward-facing steps that follow a straight line, but rather by a series of mind-spinning and consciousness-altering steps up the landings of a winding, spiraling staircase. At each point in life a person must embrace counter-intuition, i.e., *teshuvah*, in order to advance spiritually.[11]

Chassidut teaches us that there are two types of *teshuvah*, which correspond to the two Talmudic opinions regarding whether or not *teshuvah* is a prerequisite for redemption.[12] Rabbi Eliezer's opinion is that the future redemption is dependent upon *teshuvah*, whereas Rabbi Yehoshua is of the opinion that the future redemption does not depend on our doing *teshuvah*. Maimonides rules like Rabbi Eliezer, that redemption *is* dependent on *teshuvah*.

This Talmudic dispute is echoed in another dispute between Rabbi Eliezer and Rabbi Yehoshua,[13] in which Rabbi Eliezer is of the opinion that the future redemption will take place in the month of *Tishrei*, while Rabbi Yehoshua's opinion is that the month of *Nisan*—the month of our deliverance from bondage in Egypt—will be the month of the future redemption by the Mashiach.

It is thus explained in Chassidut that the *teshuvah* that Maimonides is referring to is a particular type of *teshuvah* that is uniquely relevant to the month of *Nisan*. This type of *teshuvah* is different from the *teshuvah* that we do during the month of Elul and the High Holy Days in the month of *Tishrei*. The *teshuvah* of *Tishrei* is based on strengthening (*hitchazkut*) and improvement (*hishtaprut*); we try to rectify ourselves based on what we know to be wrong in our lives. A person makes an account of his actions and knows that there are things that he should be doing differently. The normal sense of *teshuvah* is to look at myself and ask whether what I am doing is right, and then deciding to improve based on my knowledge of what is right.

But, the *teshuvah* of *Nisan* is based on renewal (*hitchadshut*). In *Nisan* we seek a new life, a complete metamorphosis, not just a rectification of the past; we are looking for a new mindset. In *Nisan*, I come to the revolutionary realization that I have never truly known what is right and what is wrong.

Similarly, the *ba'al teshuvah* of today is completely different from the *ba'al teshuvah* of past centuries. A hundred or two-hundred years ago, a Jewish individual would have had a basic Jewish education that imbued him with the knowledge of what constitutes proper and improper behavior. Equipped with this knowledge, he would have been aware of his misdeeds and could decide to improve his actions through *teshuvah*. But, in our generation, *teshuvah* often entails actually being reborn and taking on an entirely new understanding of the world. Indeed, seeing things from a completely different perspective is the true essence of *teshuvah*. This is the type of *teshuvah* that is required to bring the Mashiach.

As the sages state: "In [the month of] *Nisan* we were redeemed, and [in the month of] *Nisan* we will be redeemed." Clearly, this means that the redemption is a counter-intuitive process. The *teshuvah* necessary for redemption is a counter-intuitive type of *teshuvah*. It requires us to rethink everything that we know, our entire perspective on life. I have to understand that my whole perspective on life was wrong. This is the *teshuvah* of Mashiach. So that is our first *gematria*. The phrase "*ipcha mistabra*" teaches us about the essence of the true *ba'al teshuvah*.

Silence

Now let us turn our attention to a second *gematria* for 815. This time it is one word whose *gematria* equals 815, "silence" (שְׁתִיקָה). Now, at first glance, silence may not seem to reveal very much about counter-intuition, *ipcha mistabra*. So, to help us understand the relationship between the two, let us start by quoting the sages' saying about the function of silence: "The fence around wisdom is silence."[14] Since in Torah, wisdom refers to the ability to intuit correctly, according to the sages, it follows from this adage that if a person wants to arrive at true intuition, he has to be totally quiet, totally silent. Silence is the fence that surrounds true intuition.

In Chassidut, silence is further related to the word "*chash*" (חָשׁ), the first stage in the three stage process taught by the Ba'al Shem Tov as the cornerstone of all spiritual transformation.[15] Silence allows you to give up your previous understanding in order to ascend to a higher level of consciousness.

When Einstein first taught his special theory of relativity, some older scientists said that it looked good on paper, but that they were too old to get into it and adopt its new way of thinking about the universe. In other words, they were already too entrenched in the old perception of the universe and could not counter their common-sense to adopt Einstein's new intuitions.[16] Given that Einstein's theory turned out to be a tremendous success, these scientists would have agreed that the moral is that you should never become too old to adopt new ways of thinking.

How can we retain our minds' youthful virility? By making use of silence—the instrument for embracing counter-intuition! Silence allows us to give up our old way of thinking about the world and accept a new one.

Counter-intuition is the most important faculty that modern science has taught us to exercise. Scientists have to be willing to make existential leaps (existential because for a scientist the paradigm by which his common-sense functions is the essence of his self). They have to be willing and able not just to surrender their common-sense, but to also be willing to abandon what everyone else thinks is true. Uprooting the paradigms of their own intuitions was a prerequisite for scientists to transcend and go beyond the accepted Newtonian and Hamiltonian mechanics of their time. And as it turns out, without these leaps they would not have been able to advance our understanding of nature.

This is the spiritual moral of the scientific story of the last few generations. We have to open our minds, to be willing to change our mind-set in order to open up to a new spring, a new birth, which vis à vis everything we have thought until now, is entirely counter-intuitive.

Now, let us turn to the three great theories of the 20th century—special relativity, general relativity, and quantum mechanics—and see what aspect of our common-sense each challenges. But first, let us note an important distinction between these three theories and the relative newcomer, string theory, in relation to counter-intuition.

Counter-intuition in physical theories

Both special and general relativity were developed by Einstein in the first years of the 20th century. Though their names suggest they are two developmental stages of the same theory, to this day, scientists consider them to be two distinct theories. Special relativity was introduced into scientific consciousness in 1905, the year Einstein presented three ground-breaking papers that literally turned the world of physics upside-down.

Quantum mechanics, on the other hand, was a group effort, developed by many scientists who collaborated over the first quarter of the 20th century.

Finally, string theory was first explored in the late 60s and early 70s, but did not really take-off until the 90s. Like quantum mechanics, string theory is the product of collaboration between many scientists.

But, what string theory clearly lacks is the same degree of counter-intuitive challenge to our common sense that special and general relativity and quantum mechanics pose. String theory's challenge to our common-sense is quite weak, in spite of the fact that it contains strange sounding ideas like the universe having 10 or 11 or 26 dimensions. It can be compared to a gentle giant that has the power to lift us up high, but in a calm and non-disruptive fashion. The excitement it has generated is not because of its counter-intuitive principles, but more because of its ability to unify the already counter-intuitive realities described by relativity and quantum mechanics.

It is important to note that the first three theories are pretty much accepted across-the-board by scientists, whereas string theory is not. A minor reason for string theory's relatively weak standing is its relative youth, but the major reason is that string theory is currently unable to produce testable predictions. Indeed, the criticism of string theory does not focus on how counter-intuitive it is, but on its inability to produce testable predictions.

Let us now look at the three main physical theories of the twentieth century and analyze each one's particular attack on our common-sense, both scientifically and in the language of Kabbalah. Clearly, each of these theories is complex and can be described in many different ways. We will try to focus our attention on each theory's most surprising counter-intuitive component.

In order to help us orient ourselves, before we begin we will present a bird's-eye view of what follows. Special relativity is counter-intuitive to the mind-set of the World of Action, general relativity is counter-intuitive to the common-sense of the World of Formation, and quantum mechanics challenges the common-sense of the World of Creation.

theory	challenges mindset of World of
quantum mechanics	Creation
general relativity	Formation
special relativity	Action

The counter-intuitive aspect of special relativity

In the World of Action, the lowest of the three Worlds that our consciousness inhabits, we sense that we are moving; we are advancing in life. The most intuitive thing about the World of Action is that we think that everything is progressing. If someone would tell you that you are actually standing absolutely still, that would be totally counter-intuitive. At this level, our consciousness feels that everybody and everything is active. Everyone is on the move, going to work, doing something.

Enter special relativity, which claims that our sense of motion is really just a subjective illusion that our mind (our consciousness in the World of Action) tells us is happening. Obviously this is really counter-intuitive.

What is special relativity's argument? It starts from the basic idea that motion is always relative. We judge our motion in relation to another physical body we deem to be at rest. But, in fact, even though that other physical body seems to be at rest, it is only at rest relative to us. But, relative to a third object, it is actually in motion. If we continue to follow this chain of relative motion all the way, we find that there is only one thing that is constantly, absolutely moving in respect to everything else. That one thing is light.

Special relativity challenges our basic intuition about the role that space and time play in everyday action and events. Not only does special relativity merge space and time into a single entity (meaning, that from the standpoint of relativity you are always in relative motion, whether it be through space or through time), but it also casts them in a different role.

Until Einstein developed special relativity, our intuition told us that space and time were both constant and absolute, meaning that all observers should measure distances and time in the same way. In other words, 5 hours really do not take any longer whether you are on the red-eye flight from Los Angeles to New York, or whether you are sound asleep in your bedroom. This was the way Newton understood space and time. His (and our everyday) intuition is that space and time serve as the backdrop to events and actions that happen within their framework and context. Space and time neither contribute to nor change because of the events that take place in them.

But, special relativity challenges this so common part of our common-sense. Einstein's innovation was that space and time are not a context at all, but that they actively participate and are affected by events. If you are traveling at a very high velocity time dilates and space gets shorter (the Lorentz effect). Time and space are variable, not constant.

In his theory of special relativity, Einstein hypothesized that the only constant and absolute motion in nature is that of light. It does not matter how fast you are going, light will always be moving away or towards you at exactly the same speed: about 300,000 kilometers per second. Everything else, including space-time itself, is relative to the observer. This means that while our common-sense treats space and time as purely objective entities, special relativity makes them purely subjective. This is another way of rephrasing how special relativity challenges the common-sense mind-set of the World of Action.

Incidentally, by challenging our common-sense notion of space and time as constant and objective, special relativity requires us to reframe (and perhaps even discard) many seeming contradictions between scientific theory and the Torah. For example, consider the age of the universe. Since we can no longer refer to some absolute reference for the passage of time, we have to get used to the strange counter-intuitive notion that time can move faster or slower depending on our motion. Indeed, if we measure time from the point of view of a beam of light, there is no passage of time whatsoever. According to special relativity, without unambiguously defining "the observer" in the Torah's account of creation and the observer in the scientific theory of creation, one cannot compare or contrast the two, because they lack a common frame of reference.

As explained earlier in this lecture, the three major theories of modern physics imbue a higher understanding into our lower consciousness. The higher understanding formulates as the counter-intuitive elements in the theory, while our common-sense attitude towards reality represents the lower consciousness that these elements are challenging. The source of this higher understanding, in the language of Kabbalah, is in the World of Emanation. In a certain sense, we can say that these theories not only elevate our consciousness of the universe and our place in it, but more importantly, they change how we imagine our relationship with God.

The role of mathematics in countering common-sense

We will see how this is in a moment. But, the question first has to be asked: What is the true nature of reality? Is it what our common-sense tells us it is? Or, is it what Einstein and his theory of relativity tell us it is? This is also related to another question: How is it possible that Einstein and other scientists so intuited such counter-intuitive notions?

Above we used the image of a ray of light descending from the World of Emanation and permeating our lower consciousness. Now let us add that the Magid of Mezritch, the Ba'al Shem Tov's successor, was fond of saying (in Yiddish) that "*Atzilus is auch da*!" meaning that, "Emanation is also here!" What the Magid meant is that there really is only one reality and it is only different modes of consciousness that cause us to see it in different ways. The mode of consciousness of Emanation depicts reality in the brightest light, without severing it from the Creator. So, if you have attained the consciousness of Emanation, you will see that it is here too. But, if you have not yet, then you can only catch a glimpse of reality through the eyes of Emanation, whenever Emanation-consciousness descends into your own mind-frame.

When it comes to understanding the universe's mechanics, it is with the aid of mathematics that we can attain the consciousness of Emanation. For this reason all three theories (quantum mechanics and general and special relativity), even though they counter our experiential intuition, were discovered using mathematical considerations and reasoning. When it comes to physical laws, it is mathematics that carries the consciousness of Emanation down into our consciousness. Mathematics is like the ray of light that brings the consciousness of Emanation down into the three lower Worlds.

The value of challenging our common-sense

At this point, we have to take a step back and ask: What is the value of having lower consciousness in the first place? Why did God not just create our minds and senses in a way that they would experience reality as special relativity says it is? Clearly, this is not a scientific question but a religious one requiring us to find an answer in the Torah. So, let us rephrase this question by going back to the idiom *ipcha mistabra*, which refers to counter-intuition.

As noted, the Jerusalem Talmud makes no use of this idiom because, as we explained, its intuition is correct from the outset and does not require challenges. But, the Babylonian Talmud does, implying that its intuitive understanding is sometimes mistaken and requires corrections attained through the discourse between the sages. The sages described their state of consciousness in Babylon as that of one who dwells in darkness.[17] Like a person trying to judge what objects are in the dark, their initial intuition may be incorrect, until the lights are turned on, so to speak. So which is better? Is it better to always have higher intuition, like the Jerusalem Talmud, or is it better to first err and then be corrected by the light of higher understanding?

The essence of this question is disputed in the Talmud in a different context.[18] There is a question of who is greater: the *tzadik*, who never transgresses from the start, or one who transgressed and then did *teshuvah* (repented)? One opinion is that the *tzadik* is always greater than the *ba'al teshuvah* (Maimonides does not rule this way in his code of law). Following the logic of this opinion, the Jerusalem Talmud, which never needs to have its intuition challenged, is greater than the Babylonian Talmud, which often has to do exactly that. But, according to a second opinion, it is the *ba'al teshuvah* who is greater than the consummate *tzadik*. The *ba'al teshuvah* has arrived at the correct conclusion after having spent time living with mistaken intuitions about life. Indeed, his experience has led him to a more profound understanding regarding the depth of the Torah's advice than the *tzadik* who has never overstepped the boundaries of the Torah. Maimonides and others rule according to this second opinion, which when applied to our question implies that countering one's intuition and forging a corrected common-sense is in the end more rewarding and revealing than having the right intuition from the outset.

In Chassidic teachings a similar idea is stated with regard to light: light that comes out of darkness is greater than light that did not come from darkness.[19] In retrospect, it is a positive thing that your initial common-sense was wrong and that you spent some time in relative darkness, as long as you had enough courage to silence your own mind and counter your intuition when it became clear that this was the right thing to do. In that case, the entire journey was worthwhile, because the light that comes from the metamorphosis of one's heart and mind is the light of *teshuvah*, which allows the *ba'al teshuvah* to attain levels that even a perfect *tzadik* cannot. So this is an example of how the World of Emanation illuminates the relative darkness of each of the three lower Worlds.

Let us restate what we have seen:

Special relativity is a total reorientation of our approach towards the universe. It does two things: it connects and unites matter and energy, by the most famous equation in all of science: $E = mc^2$. And, not only does it do that, but it also makes space and time part of the events and not just the backdrop in which they occur. This means that space and time are relative concepts, relative to the observer, while the only thing that is absolute is the speed of light.

Interestingly, one of the most recent scientific developments is that perhaps even the speed of light itself, Einstein's one and only absolute, changes with time. Scientists conjecture that at the beginning of creation, the speed of light was much faster than it is now. This is an idea that is gaining popularity, even though it still cannot be proven. In a similar vein, scientists have hypothesized that the fine-structure constant may also have changed.

These new discoveries could also explain many things in accordance with the Torah. Indeed, the Rebbe taught us that none of the laws of nature should be assumed to be necessarily constant. What this means is that contrary to the assumption that nature is and has always been uniform, we cannot extrapolate about the speed of light in the past based on our current measurements.

To summarize, special relativity states that everything depends on the observer, but this is counter-intuitive to the normal mind-set of the World of Action. What this boils down to is that a person could be running his entire life and in the end realize that he has gone nowhere. The only one who is really going anywhere is one who is attached to light. "Light is Torah."[20] If you are on a beam of light you are moving. This is absolutely counter-intuitive to the consciousness of the World of *Asiyah*.

The counter-intuitive aspect of general relativity

Now what about the World of Formation (*Yetzirah*)?

General relativity takes into consideration gravity, one of the four fundamental forces. It is the most elusive of the forces. But what Einstein realized is that from an observer's perspective, the experience of acceleration and the experience of gravity's pull are equivalent. This has very far-reaching consequences. The example that is usually given to demonstrate this is that if you are traveling up in an elevator, when the elevator accelerates you feel yourself being pulled into the floor. The experience of being pulled down is identical to what you would feel if you were suddenly affected by a gravitational field. This insight forms the basis for general relativity and is known as the equivalence principle.

The most important consequence of the equivalence principle is that space is not flat, or in other words, it cannot be described using Euclidean geometry. Until Einstein, everybody envisioned space as being flat. Thanks to Riemann's work, Einstein had the necessary mathematical tools (or "vessels"[21] to use Kabbalistic terminology) to consider that space might be curved, which means that everything indeed moves in a straight line, while at the same time following the curvature of space created by the masses lying in it. Gravity, as Einstein realized, actually curves space. The idea that space is curved is counter-intuitive to what we normally conceive its geometry to be like. After having studied geometry in high school, general relativity requires us to re-conceive what space is like. Instead of space being flat it is either convex or concave.

What does this symbolize, that you take something that you thought was straight and now it is either convex or concave? In Hebrew, this is described by the idiom, "Like clay in the hands of the artist" (כַּחֹמֶר בְּיַד הַיּוֹצֵר). This is a deep insight into what the word "formation" means. The artisan has a potter's wheel, on which he places the raw material, the clay, and he curves it. In *Halachah*, a vessel that has no curvature, meaning it does not have an inner space, convex or concave in which it can hold something, such a vessel cannot be defiled (it cannot become ritually impure), thus it is not a vessel at all. The idea that space-time itself has intrinsic curvature is the absolute formative quality in the World of Formation. So what is being revealed here is that space is a malleable material, not at all what we thought it was. Counter-intuition in the World of Formation challenges what formation is really about: curvature (of space-time). Similarly, we may say that in the World of Action the one true constant of nature, the speed of light, is the essence of action, what action really is. In Chassidic thought a good deed (an action, performed in the World of Action) is only one that shines light in the world.

The counter-intuitive aspect of quantum mechanics

The counter-intuitive principle introduced by quantum mechanics is Heisenberg's famous "uncertainty principle." What it started out as is that we cannot know both the location and the momentum of an elementary particle at the same time. At first people thought this was a limitation introduced by the observer. But, later it became clear that quantum mechanics is saying something far more profound. It is telling us that in truth, in essence, a particle does not have exact position and exact momentum at the same time. Uncertainty is not the result of limited powers of observation, rather it is built into the very concept, the very nature of the particle. This means that particles are no longer "things," as Dirac said. Elementary particles, like electrons, are not physical objects, they are wave functions, meaning probability functions. There are many ways to mathematically formalize this.

Feynman, one of the greatest American Jewish physicists used to say that there are perhaps a few people who understand, who grasp, what relativity is saying. But, there is no one at all who can fathom the meaning behind quantum mechanics.

Let us for a moment restate what the counter-intuitions of the Worlds of Action and Formation say about our service of God. In the World of Action, we said that the counter-intuition boils down to changing our comprehension of what motion is, what it means to be in motion. This is what the prophet says that relative to everything else a person is a walker, a mover.[22] Relative to everything else, only a human being possessing a Divine soul can really move because he is on a beam of light (the light of the Torah).

In the World of Formation, curvature is a statement about the human heart. The heart is not something flat, it is something that needs to be formed, that needs to be shaped (sometimes, it needs to be broken) in order to become a receptacle. You have to make your heart into a vessel. You have to be a craftsman, an artisan in order to make your heart into a vessel.

Uncertainty, the World of Creation's counter-intuition, is a statement about the mind. The paradox of the mind is that physical matter is intangible. In Kabbalah, it says that when consciousness reaches the World of Creation, it no longer relates to "things," there is only raw material, i.e., the potential for things. This can alternately be described as conceiving of the appearance of physical matter and events as they will evolve in the lower Worlds, but not their physical reality itself. At the level of the World of Creation, the mind has to divorce itself from thinking about "things." The mind has to reform itself completely around this idea of uncertainty which takes away our notion of things. As we noted above with regard to the World of Formation (and the World of Action), here too, in the World of Creation, counter-intuition—the recognition that things do not really exist as we experience them (only in potential or as mathematical expressions)—is the essence (from the perspective of the World of Emanation) of what Creation (*ex-nihilo*) is.

Lecture 2

Introduction

We will continue now with a famous saying of the Ba'al Shem Tov: "No matter what the *chidush* (innovation) of any of the great Rabbis (and in his time there were very great sages), I can contradict it." This is truly a remarkable statement. What was the Ba'al Shem Tov trying to say by this? Clearly, he was not trying to make himself out to be the smartest of all the sages. What does it mean that he can find a loophole in any innovation in Torah, or in any line of reasoning for that matter?

The Ba'al Shem Tov himself explained his meaning. What he meant is that every theory that is proposed by any Torah authority is a thought experiment in the realm of a particular World. But, in the middle-most point of that reality, there lies an essential nothingness. There is only one great soul—called the "Moshe Rabbeinu [Moses] of the generation"—that is in tune with this zero-point[23] that lies in the middle of every theory, and the Ba'al Shem Tov was in tune with that zero-point.

What is the purpose of the central point of nothingness? The points of nothingness in each World, at each level of consciousness, are connected to one another. Together, all the zero-points through all the Worlds are like a tube, or pillar, through which the person who is in tune with them can ascend and descend through all the Worlds. Now this is a very deep and important realization for science in general.

Every scientist, as great as he is, to the extent that he is really honest must ask the question whether science can ever really reach the end of understanding. Is science able to present a theory or equation that will completely encompass and represent all of creation? There are many audacious scientists who think that we are on the verge of formulating exactly such a description that will indeed explain everything. We will know everything with one formula. When that happens, all the scientists will have to retire, because then there will be nothing more to discover about the world. With one formula we will know the entire story of creation.

It's truly an amazing thing that some scientists, who are usually very smart people, do not learn from experience. The sages say that a wise person is one who learns from experience.[24] The experience of the last few hundred years has shown time and again that every theory has always been consumed by a successor. There has always been a small loophole left over in every theory posited, some phenomenon or data that could not be explained, the existence of which eventually led to the creation of a new, more encompassing theory, which then consumed the previous theory. And then scientists think that, "Now we have it made." But, again there is always one little loophole left that cannot be explained. And yet, every time that a new theory is developed, scientists think once more that they are on the verge of understanding it all, if only that one little thing that we can't explain would just go away. In fact, that one little thing is exactly what reveals the shortcomings of the theory. And then, some new theory comes about. So how naïve is it on the part of some scientists to think that we are on the verge of explaining everything, leaving no room for doubt and no loopholes.

This is what the Ba'al Shem Tov was saying. Whatever theory you present, because I am connected with the inner point of nothingness, I can find the loophole in it. And this is actually a good thing, because by finding the loophole, the unexplainable nothingness that lies at the core of every theory, the Ba'al Shem Tov forces you to uncover a higher theory.

We have to say that it is by Divine Providence that scientists are naïve in this respect, because, were they not, they might just give up on studying science. They might prefer to become historians or musicians or something else. But because they have this initiative and vigor and momentum that, "we are on the verge of knowing it all," they stay on their toes, constantly excited by discovery.

The Ba'al Shem Tov also did not mean that his statement cause Rabbis to stop studying Torah, or that they refrain from teaching and/or writing their understanding of the Talmud; on the contrary (*ipcha mistabra!*). He just told them that whatever deep understanding they will arrive at, he could find the loophole in it.

Now, the loophole in your understanding is called a "*stirah*," which means a "contradiction." But, the root of this word in Hebrew is the same as that for the word that means "concealed" (*hanistarot*, as in the verse "The concealed things are for *Havayah* our God"[25]).

There is another point to be made here. Rav Chaim Vital once asked his mentor, the Arizal, why it is that things that were supposed to be concealed even from the greatest Kabbalists, were now being revealed by him, the Arizal. The Arizal answered that no matter how much you reveal there will always be *nistarot*, there will always be something more that remains concealed. It will simply be at a higher level. Every revealed level has a loophole.

Applying this to my own personal life, this means that regardless of how I live, there is always a loophole in my lifestyle. There is always a concealed element, an unconscious or concealed contradiction and loophole that can serve as the gateway to ascend to a higher level. And, just as with the Ba'al Shem Tov, if I connect myself to the person who can show me this loophole, I can ascend through it to a higher level of consciousness.

In Hebrew, the word for "contradiction," *stirah*, also means "to tear down," meaning, that this little loophole will ultimately lead to the destruction of the previous theory and allow us to ascend to a higher level of understanding. But, regardless of how high we ascend, our new theory will also have its own *nistarot*, its own concealed loophole, or contradiction.

The equivalence principle revisited

Now, let's fill in a few things related to what we said in the previous lecture. General relativity is based on the equivalence principle, which says that the experience of acceleration and gravity are equivalent. Both special and general relativity are based on concepts that stem from our subjective experience. The uncertainty principle of quantum mechanics was at first thought to be a subjective experience or subjective limit, but it was then realized that it is a purely objective limit, one not imposed by the shortcomings of our subjective mind, but by the objective nature of reality. This realization led to an entirely counter-intuitive conclusion, one that contradicts all our subjective experience, i.e., that "an electron is not a thing." We do not know what it is. It is the formless matter,[26] which is the substance of the World of Creation. It is also called a *golem*, as we will explain later.

Let us turn now to the World of Emanation as it is revealed in the World of Formation in the form of the theory of general relativity. What is the verse from the Bible that best encapsulates and thus best alludes to the essence of general relativity? In the very beginning of the Song of Songs (1:4) we find the phrase, "Attract me, [and] we will run after you" (מָשְׁכֵנִי אַחֲרֶיךָ נָּרוּצָה). This is a beautiful encapsulation of the equivalence principle.

Gravity in Hebrew is called *meshichah* (מְשִׁיכָה), which means to draw, or to attract. As we will go on to explain, gravity is the only one of nature's forces that is purely attractive. If we would compare it with the electromagnetic force, we would say that its "charge" is mass and it is all positive. So even though it is very weak relative to the other three forces in nature, it has the greatest effect on the universe. The first word of this phrase means, "Attract me [to you]," or "Make me gravitate [towards you]" (מָשְׁכֵנִי), and then the next two words mean, "We will run after you" (אַחֲרֶיךָ נָּרוּצָה). What is the second half of the phrase alluding to? It is actually describing acceleration. This short, 3-word phrase, contains a beautiful statement of Einstein's equivalence principle, the equivalence of gravity and acceleration.

An innovative distinction between gravitation and acceleration

Now, there is a slightly irregular grammatical feature in this phrase. The first half is in the singular ("Attract *me*"), while the second half is in the plural ("And *we* will run after you..."). In Chassidic writings, it is explained that attraction (the force of gravity) affects the Divine soul alone, while running (acceleration) is felt by both the Divine and animal souls together.[27] This is certainly something more than what Einstein knew. There is definitely an equivalence in experience between gravity and acceleration. But, the experience of acceleration is shared by both the Divine and animal souls, while the initial gravitation is an arousal that God arouses the Divine soul alone with. The "gravitational pull" attracting the Divine soul to God is Divinity attracting Divinity. The soul is a part of the Divine, so the part gravitates to the whole, and the whole draws the part towards it. In this parable, Divinity is symbolized by mass, and the soul gravitating to God is like a stone falling to earth. This relates to the

Ba'al Shem Tov's famous saying that when it comes to "the essence," when you hold on to a part of it, you are actually holding on to all of it.[28] That is what spontaneous gravity is like.

But, the second part, "We will run after you," describing acceleration, means that both the Divine and animal souls experience acceleration towards the Divine. The reason for this is that acceleration is felt as something that "I am doing," which in order to be experienced this way requires that the animal soul, which can simply be described as "my feeling of self," also be involved. In any case, we find from this verse that indeed there is equivalence between the experience of gravity and the experience of acceleration.

This is another example of our method of searching for models that can illustrate scientific ideas in a verse, or in a *gematria*, etc. It is most important to find an *asmachta*, an allusion in the Torah for scientific concepts and ideas.

Energy, matter, time, and space

Now let us go back to the special theory of relativity. The famous equation here is $E = mc^2$. We will need to investigate this equation in much more depth later. What this equation does is it equates mass, or matter (denoted by the letter m), with energy (denoted by the letter E). We have here matter, which in Hebrew is called *chomer* (חֹמֶר). Energy is called *ko'ach* (כֹּחַ). So Einstein's famous equation means that matter is just another form of energy. But, every little drop of matter has a tremendous amount of energy in it, since it is multiplied by the square-value of the speed of light, a very large number.

Because of this equation, for some mysterious reason, space and time now become affected by events and by motion that takes place in them, the essential discovery of special relativity. So now we actually have four different concepts grouped in two pairs: energy and matter, and time and space.

In Kabbalah,[29] these four concepts: time, space, force, and matter correspond to the four letters of God's essential Name,[30] as follows:

Time corresponds to the letter *yud*, or *chochmah* (wisdom). For instance, it is said, "Who is wise? One who sees the future."[31] Intuition about time has to do with wisdom.

Space is a *binah* (understanding) concept, and there is a verse that states this explicitly, "Where is the place of understanding?"[32] This verse equates space and *binah* (understanding); meaning that the experience of space is a mother experience, while the experience of time is a more fatherly experience. How can we explain this in simple terms? The womanly sense is to know where everything belongs in the home, everything has its proper place. This is a sense of intuition regarding space. There are modern psychological systems, especially geared towards women, that build on this form of intuition. Womanly intuition is that everything has its proper place. To have your child grow-up with a consciousness that is based on Torah study, that is the father's responsibility. This includes planning the future, the child's career. This is a more time-oriented consciousness. The two together, wisdom and understanding, the father and mother principles, which here correspond to time and space, are called in Kabbalah: "Two partners that never sever from one another."[33]

So what Einstein did is to connect time and space into one unit, one entity; together, they are one thing. In the terminology of Kabbalah, this is called "the higher unification" (*yichuda ila'ah*). Einstein also put together energy and matter, which corresponds to the unification of the bottom two letters in God's essential Name, *vav* and *hei*. This is called "the lower unification" (*yichuda tata'ah*). Einstein had unified space and time corresponding to the unification of the first two letters in God's essential Name, *yud* and *hei*. Then he unified energy and matter, corresponding to the two final letters in God's essential Name, *vav* and *hei*.

But, then came the truly innovative result: that space and time, which before were viewed as, let us call it "transcendent"—taking no part in the exchanges of energy and matter—were now understood to be affected by physical events involving matter and energy. In Kabbalah and Chassidut, this corresponds to the manner in which the lower unification affects the higher unification. This is called the drawing of *mochin*, or brain-power, or life-force from the father and mother principles, which correspond to time and space as they are unified together, into the lower levels of energy and matter, which correspond to the lower unification.[34] Space and time change as a result of events that take place in the realm of energy and matter, which have now become one.

Now we will see something interesting about these four words (time, space, energy, and matter) as they are written in Hebrew: זְמָן מָקוֹם כֹּחַ חֹמֶר[35]. If we look at the middle letters of these four words we see that they are: זְמָן מָקוֹם כֹּחַ חֹמֶר. Just the middle letters together spell the word מָקוֹם, "space." What this reveals is that it is specifically space (which, as we mentioned, corresponds to the mother principle, the first *hei* in God's essential Name) that unites all four of these concepts. This is similar to what the Ba'al Shem Tov said that there is an inner point of zero, of nothingness within everything, and if you connect to it, you can be drawn up to a higher level. Indeed, more than the word for "time" (זְמָן), the word for "place" (מָקוֹם), in Hebrew, is a connotation for the Almighty. For example as we say in the Passover Hagaddah, "Blessed is the Place [God], blessed is He" (בָּרוּךְ הַמָּקוֹם בָּרוּךְ הוּא). The word for time, *z'man* (זְמָן), has many Kabbalistic gematrias and explanations, nonetheless, it is not a connotation for God. So there is something about space that is more Divine, more primordial, than even time. Still, Kabbalah teaches that time is more internal than space. It is time that enters space, which is more primordial. If there were space without time, it would be a continual present; there would be no flow of time.

Actually, from a ray of light's perspective, there is only space, there is no time. There is a famous explanation of this idea that says that a ray of light that leaves the sun takes 8 and a half minutes to reach Earth. Light is not said to travel with infinite velocity, but it is said that light never left the sun. But it did leave the sun. Let us explain it like this: if we have a ray of light that left some far away star so many light years ago, it has left. But, we say that the ray of light is really still in that star, it is still clinging to its source, to the star from which it emerged. Why is this so? Because from the perspective of the ray of light there is no time! Even though the light has definitely spread into space—it has traversed space—nonetheless, it is still clinging to the star. If the light would experience time, it would have left the star. But since there is no experience of time, there is no time-lapse; no time has passed from when it left to when it reaches its destination, so it is instantaneously both at its starting point and at its ending point. The light is still clinging to its source. This is a very important concept in Chassidic teachings which is referred to by the idiom "light clings to its source."[36] This most important principle in Chassidic teachings, which states that no matter what, light always clings to its source, is the Torah equivalent of the special theory of relativity. What this itself means is that there is space without time.

Once more, space is in a certain sense more essential than time. When time enters space, it becomes the male presence within the female. But space is more primordial than time, which is why it is a connotation for the Almighty. And, what happens in special relativity is that both space and time together become the *mochin*, part of the experience or the reality of *ze'er anpin*, the "Small Countenance," corresponding to the emotive attributes of the soul, to the letter *vav* of God's essential Name, and to the concept of energy in nature.

Every theory has a black hole

Let us go back to the Ba'al Shem Tov's saying that no matter what theory you might come up with, I can always find a loophole in it. The loophole in a theory can be likened to a very important idea that comes out of the general theory of relativity—the black hole.

A black hole cannot be observed, by definition. Nonetheless, most scientists believe they exist. It is even theorized that at the center of our own galaxy, the Milky Way, there is a gigantic black hole that is 2.5 million times more massive than our own sun.

Some scientists even claim that a black hole creates a new universe. A black hole is a parent of a new universe. This is exactly what the Ba'al Shem Tov said. To say that a theory has an internal contradiction is at most a philosophical statement. Now we are saying that there is an actual, a physical "nothing" that is theorized to exist at the center of our physical reality. So the physical black hole at the middle of a galaxy is like the potential nothing that creates a new something. This is just like how the loophole in every theory is the point of origin for the next theory.

We said that some scientists do not learn from experience, they are naïve, not wise. Had they learnt from experience they would recall that every theory has fallen to give rise to the next one. Why do some scientists not understand this? This idea is even more pronounced in mathematics. In mathematics, which is far more theoretical than science, this has been proven, in what is called Gödel's theorem. This theorem states that given any axiomatic mathematical system there is always an inconsistency, a mathematical black hole, if you will.[37] So if Gödel already proved that every mathematical structure has its own black hole, why not do the same for science? Why not prove that every scientific theory has a black hole by definition. Meaning, that not only would we be learning from experience that every theory has a black hole, but also prove that this is indeed the case. It is one thing to learn from experience that you will always be wrong. The next stage is not just to learn this, but to prove that this is the case.

Once more, the black hole is the physical analogy to the Ba'al Shem Tov's loophole. In string theory a galaxy and a photon are considered equivalent, something that we will reflect on later. Everything has its black hole, even the smallest thing has a point of nothingness inside.

So we have our black hole analogy, and our saying from the Ba'al Shem Tov and we have Gödel's proof, which was definitely, without question, the most important mathematical discovery of the 20th century.

In the end, what this all boils down to is that there is only one way out which is akin to the exodus from Egypt—i.e., riding God's beam of light, which makes us into real movers amongst all those standing.

Lecture 3

Symmetry in the Torah

We have in the past explained a number of times that the word for "symmetry" in the Torah is חֵן, as in the verse usually translated as, "And Noah found favor in God's eyes"[38] (וְנֹחַ מָצָא חֵן בְּעֵינֵי הוי'). If you take the mirror image of the name "Noah" in Hebrew (נֹחַ) you get the word *chen* (חֵן), which is the Hebrew word for "symmetry." So, we might say that Noah was the first symmetry physicist and he found favor in the eyes of God because he contemplated the mirror image of his own name, symmetry.

Physical symmetries

In modern science there are three basic symmetries. First let us explain what symmetry means in physics. In physics symmetry is defined more broadly than the way it is in art. In physics it means that if a system is changed in a certain way, but it still appears to be the same, then it exhibits symmetry under that change. For example, if you have a ball and you rotate the ball, the ball will still look the same, so the ball is said to be symmetric under rotation. So symmetry at its heart says that if it looks the same, it is the same. All types of symmetry, like rotational and angular, all follow this principle.

The three basic symmetries that modern science addresses are called CPT, which stands for Charge, Parity, and Time. As we will explain, the order is not correct, but this is the order in which they always appear in physics. We will explain that time is the highest, then comes charge, and the lowest is parity. They correspond to a threefold division of the *sefirot* into: intellectual, emotional, and behavioral. In Kabbalistic language, each group of *sefirot* is denoted with a different acronym: the intellectual *sefirot* are called *chabad*, which stands for the initials of the Hebrew words for wisdom, understanding, and knowledge, the intellectual faculties; the emotional faculties are called *chagat*, which stands for the initials of the Hebrew words for loving-kindness, might, and beauty; and, the behavioral group of *sefirot* are called *nehi*, which stands for the initials of the Hebrew words for victory, acknowledgment, and foundation.

Parity, charge, and time

Parity means that if you look at the world through a mirror you will see the same thing. So the mirror image being the same means that the world is equivalent under a parity transformation. Another way of saying this is that left and right can be interchanged.

Symmetry of charge means that if you change every electrical charge in the universe to its opposite, you would still have the same universe. Once more, symmetry means that if you change something and it appears to be the same, then it is essentially the same. If you change the charge of every electron into a positron and every positron into an electron, then in essence you will have changed nothing, because everything appears to still be the same.

Time symmetry is the most difficult for us to comprehend because it says that if I change time such that the future is the past and the past is the future, then nothing really has changed. This is the most counter-intuitive symmetry, because obviously the past is not the future and the future is not the past, so how can a change in the direction of time leave the universe the same. The easiest to understand is parity, the mirror image. Charge is a little more profound. But time symmetry is the most difficult to understand.

In the language of the sages, the description for all of these reversed realities (the universe after a certain symmetrical transformation) is: "I saw an upside-down world."[39] What happens when you see the world backwards in some way? As said, time symmetry is the most difficult to imagine. So how do scientists explain this? They do so by using the concept of entropy.

The principle of entropy, the second law of thermodynamics, states that as time passes, everything becomes more and more disordered. This is not just an objective experience; it is within the "psyche" of the universe itself. What this means is that under a time reversal (the past would now be the future and the future would now be the past, meaning that we would be headed forward into the past), I would still experience the future as the past, even though it is now my future. Therefore, nothing has changed, even if I change the direction of time. The universe still appears the same; the past even though it is now the future still appears to me as the past and the future, even though it is now my past will still appear as the future. All because of entropy. I can't say that I understand what this means, but let us believe that this is the case. This is time symmetry.

Gauge symmetry

Now, general symmetry, which means changing something throughout the entire universe, is much easier to understand than if you change things only in a particular phenomenon and then look to see if there is a change in the rest of the universe. So, if I find a particular phenomenon within which I perform the change, the transformation, and yet, the universe stays the same, then this is considered a much more profound symmetry. This is called gauge symmetry and it is this type of symmetry that string theory is based on.

The most important example of gauge symmetry is in relation to the strong force, which keeps the nucleus of atoms—the protons and the neutrons in the nucleus—together. The elementary particle which acts as the messenger for the strong force is called the gluon. Gluons carry a charge that is designated in terms of a color/anti-color pairing.[40] The three "colors" in this description of the gluon's charge are red, green, and blue. Now, the particular gauge symmetry governing gluons states that if all the reds turned into greens and vice versa, even though this is a change within only one of the four forces of the universe, nonetheless, the universe would remain exactly the same. This is different from taking the entire universe and placing it in front of a mirror and seeing nothing change. Therefore, this is considered a more profound symmetry.

Symmetry broken, symmetry complete

Up until several decades ago, each of these three symmetries was considered to be complete. Parity was complete, as was charge, as was time symmetry. Then came the surprising discovery that for the weak interaction, which is responsible for phenomena like radioactivity, right and left are not the same, meaning, that the symmetry of parity does not apply to the weak interaction (force).[41] This was a great discovery because it turned over what was previously believed to be true about parity symmetry: that it is universally applicable. This is a beautiful example of how quantum mechanics breaks our intuition, which feels that there should really be no difference between right and left. It is totally counter-intuitive to us to say that there should be a difference between right and left, between looking at an experiment directly or watching it through a mirror. Once more, there are four fundamental interactions in physics, the strong force, the weak force, the electromagnetic force, and gravity. Only in regard to the weak force was it shown that parity conservation is violated.

So now we have set the ground work for understanding symmetry.

Parity: the behavioral *sefirot*

Now we would like to explain that parity symmetry (or conservation) corresponds to the behavioral aspects of the soul, which comprise the *sefirot* of victory (*netzach*), thanksgiving (*hod*), and foundation (*yesod*). Parity conservation specifically refers to the two *sefirot* of victory and thanksgiving, which are considered in Kabbalah to be like two sides of one coin.[42] They constitute the ability to take initiative to reach some goal (victory) and the perseverance to see that thing through until it is accomplished (thanksgiving). Victory and thanksgiving are the two sides of one's powers of action in the world, the "active powers" (*kochot ma'asiyim*) of the soul. They are considered so complementary that sometimes (especially in the World of Chaos) they are considered to be one and the same—one *sefirah*.[43]

The allusion in the Torah that clearly indicates that parity is related to victory (*netzach*) and thanksgiving (*hod*) is found in the Zohar in the expression, "he is in victory, she is in thanksgiving."[44] Once again, victory spiritually corresponds to the self-confidence needed to get up and take initiative and is relatively the masculine half of the picture and thanksgiving (*hod*) spiritually corresponds to the strength to persevere in one's endeavor and carry it through, the relatively feminine half of the picture. So at the level of action, these are the male and female principles in the soul.

Parity in the Tabernacle's wash basin

The allusion to this is found in the Torah reading of *Vayakhel*. The last of the vessels described (which was also the first one used by the priests serving in the Tabernacle) was the wash basin (כִּיּוֹר) and its base. The basin was made from the mirrors of women.[45] Halachically, a man is not permitted to look at himself in a mirror. A mirror is in essence a feminine object. What this means is that parity (symmetry between left and right) is in its essence a feminine principle.[46]

The word describing the mirrors that were used to make the wash basin, "*tzov'ot*" (צֹבְאֹת), literally means "hosts." *Rashi* explains this unique word. He explains that these were the mirrors used by the Jewish women in Egypt to beautify themselves for their husbands, even in the midst of slavery. The men were very fatigued and had no strength to be with their wives and have relations with them because of their hard labor and this could have spelled the end of the Jewish people. The women had these special mirrors. When the husband came home late at night they would call him and say "Let's take a look at our faces in the mirror [see our mirror image] and see who is more beautiful." In this manner, by having the husband and the wife together look in the mirror, they would awaken and arouse their husbands. From this came out the 600,000 children that were finally redeemed from the severe servitude in Egypt, the people that the Torah refers to as "*tzivot Hashem*" (the hosts of God) and went on to receive the Torah at Mt. Sinai, thus beginning Jewish history.

It is said that because mirrors are used to arouse the evil inclination with the desires of the flesh, when the women brought them as their contribution to the building of the Tabernacle, at first Moshe Rabbeinu did not want to receive them. But God told him to take those mirrors because it was due to them that the people merited being fruitful and multiplying. The entire birth of the Jewish people was thanks to these mirrors. God professed that these mirrors, out of which the wash basin was made, were the most cherished part of the Tabernacle.[47]

Just as these mirrors united husband and wife during the terrible slavery in Egypt, so in the Tabernacle and later in the Holy Temple in Jerusalem, water taken from the wash basin was used to make peace and reunite a husband with his wife who he suspected of infidelity.[48]

On a daily basis, the basin was used by the priests to sanctify their hands and feet before serving in the Tabernacle. To do so, the priest would place his right hand on his right foot and his left hand on his left foot and pour water from the basin on all four together. If he did not wash in this way, then his entire service was disqualified.

From these two instances we see that the basin is a symbol of connecting left and right together. The right hand and the right foot were unified, then the left hand and the left foot were unified, and then all four were washed together.

In Kabbalah, each of the different vessels of the Tabernacle corresponds to a particular *sefirah*. The basin and its base correspond to the *sefirot* of victory and thanksgiving. In fact the word *tzovot* (צֹבְאֹת), when pronounced a little differently (but spelled with the same letters) is God's Name, the God of Hosts (צְבָאוֹת). This particular Name corresponds to the *sefirot* of victory and thanksgiving.[49] This we refer to as mirror symmetry.

A Torah insight into parity conservation

Before, regarding our discussion of the equivalence principle (that gravity and acceleration are equivalent) we saw that the Torah offers us an additional insight. Here we also see a similar innovative insight.

When the husband and wife looked in the mirror they were the same except their position changed. It is interesting to ask, when is mirror symmetry most apparent in Jewish life? Let us think about the bride and groom standing under the *chupah*, the wedding canopy. Who stands on the right and who stands on the left? According to the prevailing custom, we place the bride to the right of the groom.[50]

Why should that be? In the *sefirot*, the feminine half of the victory-thanksgiving pair is thanksgiving and it is the left of its masculine counterpart, victory. But, relative to the Rabbi who is sanctifying the couple it is the opposite. For him the groom is to his right and the bride is to his left. This is a very simple way to explain how left and right interchange, it is still the same universe; the same two people are getting married whether I look at it from their perspective or from the perspective of the Rabbi.

Now, surprisingly, when the husband and his wife looked in the mirror together, the mirror acted. The mirror image was not passive. The mirror image was active. First of all it connected them together. The husband and the wife are sitting together on a bench and the husband is about to fall asleep because he is so tired from his work, there is no connection between them. All of a sudden the wife takes out her mirror and says to her husband, "Let's look in the mirror." And then something new happens, by seeing themselves together, the husband is aroused, etc. And once again, the Almighty said that this was the most beloved of all contributions made to the Tabernacle. So the symmetry made the husband take initiative and made him interested in his wife. This means that the symmetry awakened in him his power of victory. About the feminine it says that "understanding extends to thanksgiving." This is the perseverance to see things through mentioned earlier. And with this power of thanksgiving, of perseverance, the wife was now entrusted with the responsibility of carrying the pregnancy through to term. But, the arousal of victory was achieved by the symmetry.

So, we have seen how parity symmetry, the conservation of parity, is related to victory and thanksgiving.

Charge and the emotional *sefirot*

Let us now turn to charge symmetry. Charge symmetry is the symmetry of the heart's essential emotions. To be charged is to have an emotion, an emotion is a charge. People are charged, everyone who has emotions is charged.

Now, here is the amazing thing about emotions. Imagine two people, one who is all love and one who is all fear or awe. Now imagine that they switch. If all the lovers became fearers and all the fearers became lovers, the universe would remain exactly the same; meaning, that in a certain sense, it makes no difference. "These and these are the words of the living God."[51]

In the terminology of Kabbalah, charge symmetry is called *chasadim* (attributes of loving-kindness) and *gevurot* (attributes of judgment). The *sefirah* of knowledge is the origin of both loving-kindness and might, or fear. It is the source of both aspects of loving-kindness and aspects of judgment, which when they manifest in the heart become the two *sefirot* of loving-kindness and might, which are experienced as love and fear.

In Kabbalah, the definition of the interchange between loving-kindness and might is referred to as *achlifu duchtayhu* (אַחְלִיפוּ דּוּכְתַיהוּ)[52] and in Aramaic literally means, "they change places." What happens in such a case is that the light of loving-kindness enters the vessel of might and the light of might enters the vessel of loving-kindness. Such a cross-over is to be experienced every day when we read the first and second paragraphs of the *Shema*, as explained in the intentions of the Arizal. The first paragraph, which begins with the words "And you shall love *Havayah*, your God...," is the paragraph of love. But, the Divine Name that illuminates it is the Name of 42 letters,[53] which is associated with *gevurah*, with might, or fear. The Tzemach Tzedek, the third Lubavitcher Rebbe explains that this is because the light of *gevurah*, of might has entered into the vessel of loving-kindness. In the second paragraph, the opposite occurs, the Name of 72, associated with loving-kindness, illuminates this paragraph which in and of itself is the vessel of might.[54] So what takes place when we say the *Shema* twice every day is a charge symmetry phenomenon, which is called *achlifu duchtayhu*.

What this means for each of us individually is that we should remain open to a total metamorphosis or change, but that things will still remain symmetric. For example, if my primary emotion is love and I know someone who's primary emotion is fear, then the easiest way to give him an experience of love is if I assume a state of fear. If I do so, he will take on love. Charge symmetry will ensure that everything remains the same, but as we mentioned above, there is something better about the symmetric state then the original state.

The best example of charge symmetry taking place in the Torah is found in the story of the binding of Isaac. Abraham who was loving-kindness acted with might (overcoming his love and acting to sacrifice his beloved son), and Isaac who was might acted with loving-kindness (by lovingly accepting his father's judgment, willing to be sacrificed for God).

Charge symmetry, and good and evil

On a deeper level, we can say the same thing regarding the good and bad inclinations. It is explained in Chassidut that the good and bad inclinations (*yetzer tov* and *yetzer hara*) derive from the *chasadim* (the aspects of loving-kindness) and the *gevurot* (the aspects of judgment) of knowledge. What happens if throughout the universe they switch? This would also be charge symmetry. Inclination is an emotion, a charge, it is a drive. Just like an electron has a charge and there is a positron with an opposite charge. Understanding that the state resulting from such a switch between all the good and bad inclinations in the universe would, on the whole, leave the universe unchanged is very profound.

Very often, people think they are doing good but they are actually doing evil. Other people may think that they are doing evil, but they are really doing good. This is much more subtle. But this too is an example of charge symmetry.

(What it really boils down to is that the universe remains unchanged. Based on the verse, "And Noah found grace in the eyes of God," we can say that symmetry (grace) is the beauty of creation.)

We saw such an instance of good and evil being reversed in the previous example regarding the mirrors that made up the wash basin. Moshe Rabbeinu thought that the mirrors came from a bad inclination (they were used for sexual arousal). But, God revealed to him that the opposite was true, they were from the good inclination, used to ensure Jewish survival. Thus, the previous example of parity conservation can be used to explain charge symmetry. For Moshe Rabbeinu, the good was that there should not be sexual arousal, which he believed was driven by the evil inclination. But, God understood that the sexual arousal was actually good. This is very strange since in general we are taught that Moshe Rabbeinu sees things from God's perspective.[55] But, here we see that he sees things opposite.

As we said, regarding parity symmetry, the weak force was considered a loophole. Still scientists thought that charge symmetry held universally. But, then it was found that charge symmetry also has a loophole. Then it was thought that parity and charge symmetries hold when taken together. But, finally it was found that only when all three symmetries are taken together does universal symmetry hold. Essentially it all depends on time symmetry.

Time and the intellectual *sefirot*

So now let us explain time symmetry and how it corresponds to the intellectual faculties of the soul called *chabad*, which stands for the three *sefirot chochmah* (wisdom), *binah* (understanding), and *da'at* (knowledge).

In *Sefer Yetzirah* (the Book of Formation), it states that, "the depth of beginning" (*omek reisheet*), meaning the past, corresponds to the *sefirah* of wisdom and that, "the depth of the end" (*omek acharit*), meaning the future, corresponds to understanding. So the direction of time is from wisdom to understanding, from the past to the future. It follows then that knowledge corresponds to the present, which is in between the two. Before we said that relative to one another the origin of space is in understanding, while the origin of time is in wisdom. This still fits, since time is still in wisdom, but it is moving towards understanding. The classic example of this is that the future tense of a verb in Hebrew is formed with the 4 letters איתן,[56] which when treated as a word means "strength" or "power" (אֵיתָן). For example, "I will be, he will be, you will be, we will be" (**א**ֶהְיֶה **י**ִהְיֶה **ת**ִּהְיֶה **נ**ִהְיֶה) all examples of the future tense of the verb "to be". These letters indicate a vector force of the time coordinate from the origin of time to the end of time, which corresponds to understanding.

So, time symmetry means that if the direction of time would reverse, time would flow from understanding to wisdom, from the mother principle to the father principle. But, because of entropy, the universe would still look the same. The past would still be experienced as the past even though it is now the future, all due to the second law of thermodynamics, the law of entropy.

type of symmetry	in *sefirot*
time	intellectual
charge	emotional
parity	behavioral

A *Shema* meditation on symmetry

Now we will relate the correspondence just explored to the word "one" (אֶחָד), *echad*. We mention this word in the first verse of the *Shema*: "Hear O' Israel *Havayah* is our God, *Havayah* is one."[57] The word "one" (אֶחָד), pronounced *echad*, has three letters.

There is a basic meditation that we are taught in the *Shulchan Aruch* regarding this word, "one" (אֶחָד). The first letter, the *alef* (א, whose numerical value is 1), alludes to God's essence. Since I cannot know God's essence, there is little to meditate upon, so I say this letter quickly, without dwelling on it.[58] A bit more meditation can be done on the second letter, the *chet* (ח). However, the *mitzvah*, the requirement that we draw out the pronunciation of this word "one," specifically refers to its final letter, the *dalet* (ד).

What is the significance of these three letters? The *chet* (ח, whose numerical value is 8) represents the up-down dimension of reality, which includes 8 levels: 7 firmaments and the earth. The *dalet* (ד, whose numerical value is 4) represents the four lateral directions of the earth itself. The *alef* (א) is the root, referring to God, but also to the soul, as it is one with God, before it descends through the seven firmaments to the earth, where it is given its task of spreading the knowledge and light of God throughout the four lateral directions of the earth to reach the four corners of the earth. This is a simple meditation on the word "one," *echad*.

Just as we cannot conceive God's essence, so we cannot conceive the true unity of the soul with God and so the *alef* (א) is said quickly, without much meditation on it. The meditation truly begins with the experience of descending through the seven firmaments and the earth, by which we can then come to understand our mission in life of spreading knowledge of God through the four corners of the earth, "And you shall spread out to the west and east, north and south"[59] (וּפָרַצְתָּ יָמָּה וָקֵדְמָה צָפוֹנָה וָנֶגְבָּה). This verse literally means giving birth to many, many souls throughout the world, like the "mirrors of hosts," mentioned above. This is also the literal meaning of the verse "May your wellsprings spread out"[60] (יָפוּצוּ מַעְיְנוֹתֶיךָ חוּצָה) i.e., to have many children who will spread around the world, inhabit it, and bring God's light to it.

In any event, there are three physical dimensions, which we will refer to as up-down, right-left, and front-back. When we refer to string theory, we will see that it adds many more spatial dimensions. For some reason that we cannot explain they add only spatial dimensions, they do not add a time dimension.

The time dimension in space

Charge is not a spatial dimension. Time is certainly not a spatial dimension. Nonetheless, both time and charge project themselves upon one of the three spatial dimensions.

When considering the three spatial dimensions, **time** is reflected in depth, meaning, **front and back**. Depth is called the third dimension, like 3D which is in and out, or front and back. How do I know that time is reflected in the front-back spatial dimension? We are talking about how time unites with space and we are asking in what particular manner time unites with space. From a spiritual perspective, we say that time specifically unites with the front-back dimension of space. We know this because the Hebrew words for "front" (פָּנִים) and "back" (אָחוֹר) stem from the same roots as the words for "before" (לִפְנֵי) and "after" (אַחֲרֵי). But amazingly, the word for "before" (לִפְנֵי) referring of course to the past stems from the root that also means "front" (פָּנִים), while the word for "after" (אַחֲרֵי), referring to the future, stems from the root that also means "back" (אָחוֹר).[61] So we have here an amazing example of direction reversal similar to what we discussed before regarding the symmetry of changing the direction of the flow of time. What was the past is now the future, what was the future has now become the past.

Now, time affects the *alef* (א) in "one" (אֶחָד). What does it mean for a Jew to be moving towards the future and yet to always experience it as the future. We mentioned already that this is the most profound of symmetries. It is the most intellectual one. It means that just like God, a Jew is always looking at the Torah as the blueprint of creation, the blueprint that God used to create the world.[62] The Jew is always looking at the past, at what the Torah has to say. Every day we re-experience the giving of the Torah. As a Jew is moving forward through time, he is actually looking at the past and basing his entire life on what he knows to be the past, but somehow this is now also the future, since from his perspective he is always moving forward in time. Again, just like the Creator, by looking at the Torah—both the primordial Torah with which the world was created, and especially the Torah that was given to us at Mt. Sinai—the Jew creates a world, creating and forming his future.

Indeed, in Zachariah there is a vision where he sees the Torah as having a front and a back. It's called the flying scroll of the Torah.[63]

The parity in space

Parity, describes **left-right** symmetry, they correspond to the four lateral directions. These are the four directions of the letter (ד) of *echad*, "one" (אֶחָד). Parity again is right and left, and more generally it is the *dalet* (ד).

Let us go back for a moment. We said earlier that parity reflects the *sefirot* of *netzach* (victory) and *hod* (thanksgiving), they are called two "sides" of the same body. This is like "taking sides." *Netzach* and *hod* are two sides. Side in English derives etymologically from the word "side" (צַד) in Hebrew, pronounced *tzad*. Even though they are only two sides (right and left), they correspond to all four lateral directions. So parity corresponds to the final letter of "one" (אֶחָד), the *dalet* (ד), symbolizing the four lateral directions.

The charge in space

Spatially, **charge** corresponds to the **up-down** dimension. Scientists cannot explain why they chose these strange words, but the charge on quarks, which gives them their names, are "up" or "down," or "top" or "bottom" (this means that "charm" and "strange," the two remaining quarks, also follow this peculiar convention—charm is up and strange is down). The up-down terminology especially applies to the statement in *Sefer Yetzirah* (the Book of Formation) that, "There is no good higher than pleasure (עֹנֶג) and no evil lower than "plague" (נֶגַע)."[64]

Once more, when we discussed charge, we started with the five *chasadim* and five *gevurot*. They manifest either as love and fear, or even more extremely, as the good and evil inclinations. But really what this applies to is up-down, since the good rises all the way up to *oneg* (pleasure) and the evil descends all the way down to *nega* (plague). In any event, charge abstractly means that either you are charged "up" or "down." But, the amazing thing is that it does not matter in which way you are charged, because if the entire universe were to switch we would not be able to tell the difference. Like we said before, sometimes you have to change your own charge in order to bring out the opposite in someone else. Going back to our meditation on the word "one" in the *Shema*, we have that there is a mysterious projection of charge upon the letter *chet* (ח) in "one" (אֶחָד).

type of symmetry	spatial direction	letter in "one" (אֶחָד)
time	front-back	*alef* (א)
charge	up-down	*chet* (ח)
parity	left-right	*dalet* (ד)

The *gematria* of symmetry and unity

We will conclude this lecture with the following very beautiful expression. We saw that in the word "one" (אֶחָד), the *alef* (א) corresponds to the intellect, the *chet* (ח) to the emotions, and the *dalet* (ד) to the spreading out in actions. We also used this word to correspond the 3 symmetry categories in physics.

In Kabbalah, we are taught that every word can be written in a symmetrical form. In practical terms, pretty much the only way to illustrate symmetry is with a mirror. We have no other model for thinking about symmetry. In the *Shema* as it is written in the Torah scroll, the word "one" (אֶחָד) is written with a large letter *dalet* (ד). Now if I take the *dalet* as an axis of symmetry and write out the word "one" in both directions, I will get:

אחדחא

The value of "one" (אֶחָד) is 13. Writing it this way, the value becomes 22. Now, what will happen if we do this not in one but in three dimensions?

The value of this symmetry of "one" in 3 dimensions is easy to calculate. There is the *dalet* (ד) at the center, plus 6 pairs of the two letters, alef and chet (אח), whose value is 9. So, the total value will be 54 (6 times 9) plus the value of the central *dalet* (ד), which is 4, totalling 58. The value of the word "one" (אֶחָד), in three-dimensional symmetry is 58, exactly the value of the Hebrew word for "symmetry" (חֵן), which we have been discussing.

A much simpler phenomenon is that in reduced numbering, the value of "symmetry" (חֵן) is 13, the normative value of "one" (אֶחָד). But, as we saw, in three dimensions, "one" (אֶחָד) expands and becomes equal to 58, the value of "symmetry" (חֵן).

Lecture 4

We can certainly say that the driving force of science since Einstein has been the quest for unity. As we said before, in his theory of special relativity, Einstein unified energy and matter with space and time, by bringing space-time into the context of events that involve energy and matter, all with the equation: $E=mc^2$. But then, for the rest of his life, Einstein tried to unify gravity—as he had described it in the general theory of relativity—with electromagnetism. But, he was unsuccessful. After his death, two more forces were found that work at the sub-atomic level, the strong and the weak forces. So once more, a grand unified field theory is what everyone is looking for.

The fact that science is seeking unity is a very Jewish thing. Even though we said before that scientists are wrong in seeking a totally encompassing theory, because it can never fully be found, nonetheless, the quest for unity is the real reason that scientists continue to do so. Indeed, every step is a step forward. God makes it so that the scientists believe that they might find the ultimate theory, all in order to get them to progress and move higher and higher in their understanding.

Three methods for attaining unity

As we said, the quest for unity is a very Jewish thing, as the essence of our faith is unity. As in the verse, "Hear O' Israel, *Havayah* is our God, *Havayah* is one."[65] There are actually three different ways to attain unity. These three themselves correspond to the mind, heart, and action of the individual (i.e., the intellectual, the emotional, and the habitual faculties of the soul).

The first way to attain unity is when two entities that seem to be opposite nullify themselves to something that is above them. The classic example for this is a king who has two opposing ministers. When they are in the king's presence, because they both nullify themselves before him, they can bow down together. As soon as they rise up and go home, they return to their adversity. But, in the presence of the king they are null, and because of their nullification they become one. This is called *hitbatlut* (הִתְבַּטְּלוּת), or nullification.

There is a second way to attain unity, and this is very appropriate to our lives. Take for example a couple that wants to achieve greater unity. One way is to have common nullification to someone, for instance the Rebbe. But, then you have to be with the Rebbe (in your mind's eye) all the time, otherwise you fall back into disparity. The second approach is inter-inclusion, or *hitkalelut* (הִתְכַּלְּלוּת). In inter-inclusion each one has to reveal in himself an aspect of the other. This is something like charge symmetry—that each has the other in him.

The third approach is to undertake a project together. Even if we are different, a common objective that we are working on together will unite us. This can be thought of as cooperation (*shituf pe'ulah*, in Hebrew).

Let's try to think about these three types of unity.

Nullification is what can be achieved by the mind. But if you do not remain at a very high level of consciousness, you can readily fall from this type of unity, meaning that the mind is not everything. It's important to realize that the mind, *chabad*, is not everything, as far as unity is concerned. To this end, the Mittler Rebbe explained in *Kuntras Hahitpa'alut* (*An Essay on Excitation*) that the heart is the real essence of our service, and a person should not think that to pursue the Divine is solely an intellectual endeavor. Our purpose on earth is to rectify the heart by means of the intellect. So the mind is important, but it is not the whole objective.

To have maximum unity, one has to have all three types of unity. We have to have nullification, and inter-inclusion, and cooperation. Cooperation means that by having a common objective we will be unified. Inter-inclusion means realizing that you are in me and I am in you. Inter-inclusion allows us to exchange roles in a sense, allowing me to become you and you to become me. It's just a question of which side of my personality I manifest.

Physics pursues unity through nullification

Of all the approaches to achieve a unified field theory, before string theory, the only approach was through nullification. The model for this was extremely high temperatures at which the different forces become one. At very high temperatures, it is as if the different forces melt together, and they just become the same thing. This was demonstrated for the weak force and the electromagnetic force, which at very high temperatures are the same. Raising the temperature to such a great degree is like nullifying each until they can combine. It is a little like melting things together. So this is just nullification. At very high temperatures, even the forces of nature melt into one plasma. But, at lower temperatures they divide, which in our metaphor is similar to when the two ministers leave the king's presence.

So the theory went that at earlier times, at the very first instance of creation, at extremely high temperatures, the strong nuclear force would also combine with the weak force and electromagnetism, giving one unified force. Nonetheless, it is clear that even at higher temperatures, the gravitational force cannot be united as well. So this method is only good for the non-gravitational forces. Regardless of the height of the temperature, gravity could not be "melted" into the other three forces.

String theory and unity through cooperation

How did string theory help in this respect. String theory posits that by adding spatial dimensions, in actuality the forces merge together. Merging is like working together, like cooperation, they work together. This is what we described before as unity that corresponds to *nehi*, the habitual, action-oriented *sefirot*, *netzach* (victory), *hod* (thanksgiving), and *yesod* (foundation). Science has not yet discovered at all the intermediate level of unity through inter-inclusion. There is no theory that tries to surmise that perhaps within gravity there is some inter-inclusion of the strong or weak or the electromagnetic forces. Meaning, that there is some trace of these forces within gravity.

The future is in unity through inter-inclusion

Spiritually and psychologically these are three essential types of unity. Until science discovers a form of unity based on inter-inclusion the picture will not be complete. All the GUTs (Grand Unifying Theories) before string theory were based on melting the forces together. But, this does not work for gravity, and has not even been proven for the strong force. It has only been proven in regard to the weak and the electromagnetic forces. What remains open is unity based on inter-inclusion. On the spiritual level, inter-inclusion is the most important. By seeing that I am in you and you are in me, we can go on to do things together (as in cooperation).

Going back to our example of a married couple and their quest for unity, they first need a common commitment to one authority, like the Rebbe, then they have to find themselves in one another, and finally they have to work together. Because science does not have all three types of unity, the quest for unity is still incomplete.

The divide between general relativity and quantum mechanics

Before we conclude this lecture, let us say one more thing. The problem with 20th century science is that general relativity does not get along with quantum mechanics. At very small scales, the foam, or claustrophobia of the subatomic particles does not work with the laws of gravity which are relatively very gentle and not frantic.

General relativity says that space is curved, but it is a relatively gentle curve. There are no extreme and frantic events that affect spacetime. But, quantum mechanics claims that the smaller things get the more frantic they become. That wildness of the extremely small is caused by putting a particle, which is like a soul, into a small box, and then it goes mad. The more you try to confine an elementary particle to a determined space, the more crazy it becomes and because of that strange things happen, for example, the fabric of spacetime begins to tear apart, as well as other strange topological changes.

Relatively, general relativity is the world of rectification (*tikun*) while quantum mechanics is an example of the world of chaos (*tohu*). The two do not get along. General relativity is excellent at describing the cosmos, the very large, the macro. But, it cannot explain the very small. Quantum mechanics is just the opposite. They especially do not agree at the micro level.

This disagreement comes to the fore in special cases like a black hole, which occupies a very small space (on the scale of a particle), but has a huge mass (on the scale of many, many stars). Another place is in cosmology, at the moment of the Big Bang, when all the mass of the universe was confined to a single point. We will discuss more about this later on.

The need to unify *tikun* and *tohu*

But, what we now see is that the ultimate quest for unity revolves around getting rectification (*tikun*) and chaos (*tohu*) to unify. This is why inter-inclusion is the missing link. Because *tohu* and *tikun* will not cooperate together without inter-inclusion. They will not become good partners doing the same thing without inter-inclusion. Even at the level of nullification, which we likened to their being melted together, they will not work together on a common project.

The example of chaos and rectification in the Torah is that of Jacob, who is like general relativity and Esau, who is like quantum mechanics (as we mentioned earlier, Noah discovered special relativity. Then came Jacob and Esau). And they are not going to jive unless some inter-inclusion will be found to connect them. This is something that the Rashash teaches, that even extreme opposites like *tohu* and *tikun* do possess some inter-inclusion and that is the only way to unify them.

Unifying determinism and free-will

One last topic that we need to address is that of determinism. It is well known that Einstein preferred determinism, he did not like to think that God was playing dice with the universe. Again, Einstein did not like quantum mechanics' seeming allowance of free-will due to probability. But, the Torah tells us that free-will is one of its most basic and fundamental tenets. Quantum mechanics allows a lot more freedom. General relativity, on the other hand, follows Einstein's preference for determinism. So deep down, unifying general relativity and quantum mechanics also depends on our ability to solve the paradox of God's omniscience (determinism) versus man's free-will. General relativity is like God's omniscience. Quantum mechanics is like our freedom of choice.

Ultimately this goes back to the right and the left. Right is omniscience, while the left represents free-will. The angel of Esau is called the *samech mem* (*alef lamed*), which is a variant spelling of the Hebrew word for "left" (שְׂמֹאל). In the weak interaction (force) the left remains the left, it does not interchange with the right. In other words, because the weak force does not preserve parity, the *samech mem*, the left there, remains the *samech mem*, Esau's angel. Nonetheless, he too has to do *teshuvah*, meaning he has to change. Until, as the prophet says, finally, "Death will be swallowed up, forever,"[66] meaning that Esau's angel, the *samech mem*, will eventually be swallowed up in a black hole.

We will explore black holes and how they swallow things later. We will then also better understand how to achieve unity between quantum mechanics and general relativity.

Lecture 5

Back to symmetry

We will now continue to talk about symmetry. Symmetry is one of the most important concepts in modern science. Last time we spoke about CPT symmetry, which stands for Charge, Parity, Time. Symmetry in modern science is used in the sense that if you interchange or reverse two complementary concepts, then the end result will be the same, meaning that you will not be able to distinguish between the universe that you began with and the universe that you arrived at after your interchange.

If all the electrons in the world would suddenly change into positrons, and all the positrons would change into electrons, then you would not be able to tell the difference. Parity symmetry means that if right and left reverse, which is basically that you would be looking at the universe through a mirror, then everything would remain the same, you would not be able to tell the difference. Time symmetry is the most difficult to understand because it says that if you reversed the order of things, everything would still remain the same. Time symmetry is the most counter-intuitive of the three and it is very difficult to understand what it means.

But, as we mentioned before, none of these symmetries is complete by itself. Instead they are complete only in conjunction. Meaning, that if all the charges would reverse, and all the directions, the spatial extensions, and time would reverse, only then would you still see the same universe that we have now.

We mentioned that time relates to the mind, the intellectual level of the soul, while charge relates to the emotional attributes of the soul, and parity relates to the habitual or behavioral faculties of the soul.

Now let us turn to two additional examples of symmetry that emerge from modern physics.

Size does not matter

There is a symmetry principle, which derives solely from string theory considerations, which is the most amazing type of equivalence that can be imagined and this principle is denoted by the simple equation $R = 1/R$, where R is the radius of the universe, which is very large, millions of light years. And yet, what this equation says is that our huge universe, which is billions of light years across, is no different from a universe that is so tiny that it is inconceivably small, much smaller than even a photon. A universe with radius R is equivalent to a universe with radius 1/R. In other words, there is no difference between big and small.

"He who is large is small (and large)"

Every thought which has any truth to it must have a source in the Torah. This very last thought, has a very explicit source in the *Zohar* which reads: "He who is very small is very big, and he who is very big is very small."[67]

What this implies of course is that a person who in his own eyes[68] is very small—he is very humble—from the perspective of the Almighty, that person is actually very great. In this world, which is the world of deceit, he is small, but in the world of truth, he is very big. And the opposite is also true, a person who in this world considers himself to be very big, great, and outstanding, in the world of truth, from the true perspective of the Almighty, he is actually very small.

What string theory adds to the regular interpretation of this statement from the *Zohar*, is that both sides of the statement are true—simultaneously! The usual interpretation is that if you consider yourself to be small, in truth you are big, and if you consider yourself to be big, in truth you are small, or insignificant. But, now we are saying that to be very small is to be very big. Both are true simultaneously. So symmetry is telling me that I cannot tell the difference between the two things. Big and small are entirely equivalent. It is not just a question of perspective; they really are exactly the same. Again, this is the most counterintuitive result of modern physics. We usually think that we can tell the difference between big and small. Yet, here comes string theory and teaches us that we really cannot tell the difference between the two.

So we have just presented in short a symmetry principle, an equivalence principle, which is entirely different from CPT. It is much more than CPT symmetry.

Super-symmetry

Let us turn to super-symmetry. What super-symmetry says is that fermions and bosons can be interchanged. In this lecture we will begin talking about all the different types of elementary particles and what they have to teach us about our lives.

In general, in physics today we recognize two types of elementary particles: real particles and virtual particles. Real elementary particles divide into two categories: the *fermions* are the matter particles, and the *bosons* are the messenger, or force particles, which cause interactions. Examples of fermions are electrons and quarks, which make up the protons and the neutrons out of which matter is made.

The simple difference between bosons and fermions is in their spin. Elementary particles that have a half (or multiple of a half) spin are matter particles, or fermions. Elementary particles that have a whole (or a multiple of a whole number) spin are messenger particles, or bosons.

The concept of "whole" and "half" is straight out of the teachings of the 12th century Kabbalist, Rabbi Avraham Abulafia. Abulafia explains that the world has to simultaneously have both whole and half. In Kabbalistic terminology, this refers to the masculine and feminine aspects of reality, where the "whole" represents the masculine and the "half" represents the feminine. In modern physics, these two aspects, whole and half, appear in the context of the spin of an elementary particle.

Spin

What is particle spin? You can think of spin as being the number of times that you have to rotate a particle in order to bring it back to its starting position, so essentially spin is a symmetry consideration, because we are asking, under what kind of rotation will the particle return to its initial state?

Particles can take on spin values of either half-integers (1/2, 3/2, ...) or whole integers (1, 2, ...). Matter particles (fermions) have 1/2 spin while messenger particles have whole integer spin. Strangely, this means that electrons (fermions) for instance have to be spun twice to return to their initial state. Following the Kabbalistic terminology noted above, we can relate matter particles, which have spin values of half integers, to the feminine aspect of reality and messenger particles, which have multiples of whole spin, to the masculine aspect of reality.

Super-symmetry is fairly simple to express: if you would exchange all the fermions for bosons and all the bosons for fermions, the universe would stay the same and you would not know the difference. In our spiritual language, you could say that if all the men were exchanged for women and all the women were exchanged for men, then the universe would stay the same and you would not be able to tell the difference.

Super-partners

What super-symmetry implies is that every particle has a super-partner. The partner of every elementary particle that we know of has spin exactly ½ less than that particle. This means that every matter particle has a messenger-particle partner lurking somewhere in the quantum world, but these super-partners have yet to be found.[69] The same is true for every messenger particle with whole spin. It too has a super-partner, with spin ½ less, which identifies it as a matter particle.

It is important to stress the difference between anti-particles and super-partners. These are not the same at all. Every particle has an anti-particle, which has negative the amount of electrical charge that the particle has. But, with respect to super-symmetry and super-partners we are dealing with the particles' spin and claiming that every particle has a super-partner with ½ less spin than it has. If these super-partners would be found, it would prove super-symmetry. This would not yet prove that super-string theory is correct, but it would support it.

Why is it so difficult to find these super-partners? The problem with observing these super-partners is that they are very heavy, which means that you need a lot of energy to create them and they exist for a very short amount of time. This is also a counter-intuitive notion, because we would think that the bigger something is, the easier it should be to observe it. But the bigger the elementary particle is, the less time it exists, giving you less time to find it. One of the objectives of building bigger and bigger particle accelerators is to be able to create heavier particles. One of the hopes is that in these new, more powerful accelerators we will be able to observe the conjectured super-partners of super-symmetry.

Super-symmetry and the sanctification of marriage

The Torah allusion to super-symmetry, which states that the two basic types of particles, bosons and fermions, can be interchanged, is to be found in the beginning of the second chapter of the tractate of *Kidushin*, the Talmudic tractate that deals with marriage and the sanctification of a woman by a man. The *mishnah* states that the husband can sanctify a woman in marriage either himself or by messenger, and the woman can accept the sanctification of marriage from a man either herself or by messenger. In other words, the groom can give the wedding ring (or an amount of money) to a messenger to take to a woman, who perhaps lives far away, in order to sanctify her to be betrothed to him. The messenger can then go to the woman and sanctify her on behalf of the person who sent him. The same is true of the woman. She can receive the ring directly, or she can send a messenger to receive the ring.

The Hebrew wording of the *mishnah* leaves it open to a simple misunderstanding. In the original Hebrew, the conjunction connecting "himself" with "by messenger" is what is usually understood as the "and" conjunction.[70] Nonetheless, this is not the simple, literal meaning of the *mishnah*, for sometimes the conjunctive "and" is understood to mean "either/or." So though one might conclude that both forms of sanctification—in person *and* by messenger—are needed in order to get married, but this is not the case. A man can sanctify his wife *either* himself, in person, *or* by messenger. He does not have to do both.

Virtual particles and literal readings of the Mishnah

Nonetheless, many times, a Kabbalistic interpretation can reveal how the literal meaning is also correct. In this case, the Kabbalistic interpretation would be that the *mishnah* is also alluding to our subject of matter and messenger particles.

The man sanctifying a woman symbolizes the matter particle. His messenger sent to sanctify the bride symbolizes the messenger particle. What modern quantum mechanics has revealed is that every direct interaction between particles involves both matter *and* messenger particles, which are actually virtual particles.

In order to describe this process one needs to use diagrams developed by Richard Feynman, a Jewish physicist from Caltech, and considered by some to be the greatest physicist after Einstein. Feynman diagrams reveal how whenever there is a quantum interaction, even a direct interaction like an electromagnetic interaction between two electrons, both sides actually exchange messenger photons, which are virtual particles.

Going back to the *mishnah*. First let us say that perhaps the greatest example of interaction in the Torah is that of a man sanctifying a woman in marriage. Whenever there is an interaction, at some level, there are either actual or virtual messengers that are being passed back and forth to facilitate the interaction.

Two fires

Let us see another example of this. In Hebrew, "man" is אִישׁ, and "woman" is אִשָּׁה. They share the two letters *alef* and *shin* (אש). The two remaining letters from the two words are *yud* (י) and *hei* (ה), whose values are 10 and 5, respectively. So, in terms of their *gematria*, *yud* and *hei* have a relationship of a whole and a half. The combination *yud-hei* (י־ה) is also one of God's Names. Thus, Rabbi Akiva says that if a man and a woman merit to manifest the *yud* of the man and the *hei* of the woman and to unite them into a holy Name, then the Divine Presence dwells between them. If they do not merit to manifest and unite these two letters, the two left-over letters in each word *alef* and *shin*, which together spell "fire" (אֵשׁ) will consume them. This fire is either the fire of unholy passion, or the fire of anger. If the *yud* and *hei* are united, then the fire becomes holy fire, the fire that consumes the sacrifices in the Tabernacle.

Everything the Torah describes regarding the service that takes place in the Tabernacle alludes to secrets of both creation and the secrets of the Divine chariot. So now, based on this particular reading of the *mishnah*, both the man and the woman have a messenger in addition to being able to sanctify in person. These messengers are like angels. In fact, we usually explain that everything in the world of quantum mechanics, all the elementary particles can be understood as angelic "entities." This is especially true of the force particles, which act as messengers. So, if the male and the female both have messengers, then the male fire represents both the male and his messenger, and the same goes for the woman.

In the beginning of the Song of Songs, we find the phrase: "...support me with fires"[71] (סַמְּכוּנִי בָּאֲשִׁישׁוֹת). In the Temple there were two fires that simultaneously burned on the altar and consumed the sacrifices that were placed on it. The first was a fire that came from below, meaning that it was lit by the priests. That was the feminine fire. In direct response to the fire brought from below, fire would descend from heaven. That was the masculine fire. These are the two fires that are referred to in this phrase.

So, as we said, both the man and the woman have the two letters *alef* and *shin* in common. The *shin*, the initial letter in the word "messenger" (שָׁלִיחַ), is common to both the male and female.

There is a question raised in the Talmud about what happens if the messenger has "second thoughts." I send him to sanctify a woman, but when he meets her he decides to sanctify her for himself. This creates all kinds of problems, but in quantum mechanics this corresponds to an example of super-symmetry, because the boson (the messenger) has decided to become a fermion (a matter particle). The messenger has transformed into the person sanctifying. The same is also true for the woman.

Three symmetries and three aspects of love

We have now seen three different examples of symmetry:

- CPT (Charge Parity Time)
- Size equivalency
- Super-symmetry

Each of these is totally different from the others. To give these three a model, we refer to the three parts of the second verse of the *Shema*: "And you will love God with all of your heart, with all of your soul, and with all of your might." The three aspects of the love of the Almighty are thus:

- Love with all of your heart
- Love with all of your soul
- Love with all of your might

CPT and the heart

The sages say that to love with all of one's heart means to love God with both the positive and negative inclinations. Meaning, that one has to reach in one's consciousness an equivalency, a symmetry, or parity between the left and right sides of the heart (which are explained in the Tanya to harbor the negative and positive inclinations, respectively). As explained earlier, parity symmetry must come together with time and charge symmetry. Earlier, we saw how CPT correspond with the intellectual, emotional, and habitual faculties (in the order of time, charge, and parity). Now we are saying that as a whole, since they must go together, they correspond to the intellectual, emotive, and habitual faculties within the heart in itself.

Meaning, that from whatever angle of your heart, whether it be the habitual, the emotive, or even the innermost aspect of the heart, its intellect, you should have parity between the good and evil inclinations. Practically, this means that we have to be flexible enough to change the way in which we serve God, as explained before in relation to the *Akeidah* (the Binding) of Isaac. Even if my primary way of serving God is through fear (or awe), being flexible enough to also serve through love opens up the same possibility for someone who serves out of love.

Super-symmetry and the soul

Super-symmetry corresponds to loving God with all of one's soul. Why? One's soul represents one's function or role in life.

Feynman's explanation for the paradox inherent in the quantum world was that in reality all possible paths are taken simultaneously. In order to figure out why a particle reaches a certain destination, you have to sum over all the possible paths that lead to that destination. Sometimes, there could be an infinite number of possible paths. Nonetheless, we have to take the sum over all of these possibilities in order to fully describe what happened.

Now, in life, sometimes I am trying to reach some goal and then all of a sudden I realize that perhaps I have to take the completely opposite direction to get to that destination. CPT symmetries mean that possible paths that involve opposite directions along parity, charge, and time also have to be summed in order to get the final outcome. Sometimes I have to change my left with my right to reach the goal, or change charge, or go backwards in time.

But, super-symmetry is not just a change of path to get to the same outcome; it represents a complete change in one's "goal," as it were. CPT symmetry is a change of path, but not a change of goal.

One way to see this is in Rebbe Nachman's story of the Seven Beggars.[72] People were trying to reach the Tree of Life, but each one thought that there is a different path to reach it. It was impossible to make peace between these people because each had a different idea on how to reach the goal. Rebbe Nachman explained that the difference in directions is a consequence of the different goals that they have. Each person in essence possesses characteristics that suit the goal that he or she is trying to achieve.

We see this same problem in politics. Take for instance the politics of the Jewish people today. We have many groups, each with good intentions, all striving to reach peace on earth, but they do not agree with one another, because they each have a different path. So, people will never really agree on the path to take unless they find someone who can already embody the goal that they are trying to achieve. If we see someone living already with the goal that we are looking for, someone who is already experiencing this reality, then we might be able to agree on the path. This is like the Rebbe teaching us that to get to the revelation of the Mashiach we have to first live with Mashiach, we have to embody the very goal that we are seeking. Only then can we come to agree on a common path.

But, now returning to super-symmetry, we say that it is not just about changing one's direction, one's path, but changing the goal, changing one's "mission statement." In the language of physics, are you a boson or a fermion? Super-symmetry is thus about changing one's purpose in life, one's very identity. Super-symmetry implies that if all the goals were interchanged, the world would stay the same.

To be able to change one's mission in life requires sacrifice. In a sense, it requires dying in order to become something else. To reach your super-partner you have to sacrifice your life, your very purpose. This is what is meant by the words "with all of your soul," which the sages explain means, "Even if He [God] claims [takes] your soul." The Arizal says about this that if a person has fulfilled his mission in life, the night following that day on which the function was fulfilled is very dangerous. Every night when a person is asleep, the soul goes above. On this night, it will be told that it has finished its work. The only way to remain alive and come back down is by assuming a completely new function, a completely new purpose for one's life. This is one of the simplest intentions of the verse: "My soul, I long for you at night."[73] The only way to claim my soul back is by sacrificing myself, sacrificing my identity. This is the only way to make it through that night. This is like a fermion becoming a boson. So much for the second level and aspect of loving the Almighty.

Size symmetry and one's infinite nature

The third level of love is above even self-sacrifice and is described in the verse as "with all of your might." How can something be more than sacrifice? One reading that is given by the sages, which is difficult to understand, is that "with all of your might" means "with all of your money." It is sometimes more difficult for a person to give up all of his money, to give up his bank account, than it is for him to give up his life. How can we understand this?

One simple explanation is that giving up one's bank account is like giving up what one will bequeath as an inheritance to one's offspring, to coming generations. Spiritually, this is like giving up one's power to procreate, which in Chassidut is a manifestation of the power of the infinite in each person, because the power of procreation, the power to give birth and to sustain one's self through one's offspring, is an infinite power, the most powerful of all human abilities. Thus, giving up all of one's possessions, all of one's might, all of one's money, is like giving up all of one's future generations; all that one would have bequeathed to the coming generations. So giving up one's continuity is definitely more difficult than giving up one's present identity.

But, let us see another explanation. Literally, the word that we usually translate as "with [all of] your might" (מְאֹדֶךָ) does not mean "might," rather it means "very much." This is one's most extreme being. What is this "very muchness"? For many people this is indeed their bank account. This is exactly the idea behind the final form of symmetry derived directly from string theory. Being able to accept that size does not matter; that being small is exactly the same as being big. Translated into wealth this means that even if you are very wealthy it is exactly the same as being very poor.

In reference to money, the sages use a metaphor to capture the idea of the big being equivalent to the small. They state that wealth is like a wheel that revolves in the world—a wheel of fortune. Normally, the symbol of the wheel is meant to give a person the understanding that even if right now you are at the very top of the wheel and you have wealth, nonetheless, the wheel revolves and either you or your children will someday find yourselves at the bottom of the wheel. And likewise, even if now you are at the bottom of the wheel and poor, at some later time, either you or your children will find yourselves on top.

But now, what size symmetry, as derived directly from string theory, is saying is that there is no time-dependence, as it were. Being big and being small, being rich and being poor, being on top and being on bottom, they are exactly the same.[74] Not only that, they are happening simultaneously. Even if you are at the very top right now, you are really also at the very bottom, on a wheel there is no difference between top and bottom. If you are very, very big, you are also very, very small. This is the ultimate realization in pursuing love of the Almighty. If a person can reach this consciousness then that is giving up his "very muchness." Every person has an infinity to him, which is his "very muchness," which is much more difficult to give than giving up one's finite being.

"With all your heart" means being able to switch careers. Supersymmetry thus implies that we should, like the Lubavitcher Rebbe said, never retire. Always find some new goal to pursue. Even when you have finished one career, pick up a new one.

$R=1/R$ means being able to surrender one's sense of being infinite, be that infinity the infinitely small, or the infinitely great.

Lecture 6

Her path

The Book of Leviticus begins with the verse, "A man who sacrifices from among you a sacrifice to God." The sages state that the word "a man" (which in Hebrew is "Adam") alludes to the rectification of Adam, the first man. As explained in Chassidut, the word "from you" means that the sacrifice has to be from one's self. As the Alter Rebbe explains, this means that in order to properly bring a sacrifice, you first have to have the intent that you are bringing yourself (i.e., your own animal soul) and sacrificing it, and only then can you bring an actual animal sacrifice to place on the altar, where this animal then comes as a substitute for your own animal soul, which you are ready to sacrifice. This is similar to the second level of love, "with all of your soul," which we discussed in an earlier lecture.

In Kabbalah, when considering the symmetry of words or phrases, we first count the number of letters in the word or phrase. When there is an odd number of letters, there is a letter in the middle, which can be seen as the axis of symmetry, and we say that this word or phrase exhibits *masculine symmetry*. But, if there is an even number of letters, there is no middle letter that can serve as an axis of symmetry, therefore the symmetry of the word or phrase reflects around an imaginary axis of symmetry (between the two middle letters). We say then that a word or phrase comprising an even number of letters exhibits *feminine symmetry*.

The phrase out of Leviticus we just quoted has 6 words in it (אָדָם כִּי יַקְרִיב מִכֶּם קָרְבָּן לַי־הוה). Of these 6, four have an odd number of letters, so they have a middle letter. In Kabbalah, many times, when there are middle letters, we focus on these letters themselves.

The middle letter of "a man" (אָדָם) is ד.

The middle letter of "who sacrifices" (יַקְרִיב) is ר

The middle letter of "among you" (מִכֶּם) is כ.

Finally, the middle letter of "to God" (לי־הוה) is ה.

Together, these 4 letters spell the word, "her path" (דַּרְכָּהּ), where the word "way" (דֶּרֶךְ) here is in the feminine possessive form.

Wisdom's path and location

In the language of the sages, this word, "her path" appears often, especially in the context of "the path of Torah."[75] But, interestingly, this word appears only once in the entire Bible. It appears in a verse in the 28th chapter of Job. The 28th chapter of Job is such a unique and important chapter that the sages gave it a special name (there is no other chapter in the Bible that they gave a name to). They called it "the chapter of wisdom."

Wisdom obviously refers to the wisdom of creation, as another verse states: "You created everything with wisdom."[76] Even the first word of the Torah, "In the beginning," alludes to wisdom, as in the verse "The beginning of wisdom...."[77] One of the most important and famous verses from the 28th chapter of Job is, "Wisdom is found from nothing and where is the place of understanding" (וְהַחָכְמָה מֵאַיִן תִּמָּצֵא וְאֵי זֶה מְקוֹם בִּינָה). Towards the end of this same chapter we find a verse, which contains the only appearance of the word "her path" in the entire Bible. This verse is, "God [and God alone] understood *her path* [the path of wisdom—wisdom has feminine gender in Hebrew], and He [and He alone] knows her location"[78] (אֶ־לֹהִים הֵבִין דַּרְכָּהּ וְהוּא יָדַע אֶת מְקוֹמָהּ).

There is a verse that describes wisdom as one's sister, "Say to wisdom, You are my sister,"[79] meaning that a person should relate to wisdom as one relates to one's sister, in a very close and intimate way. Even a man's wife is sometimes called his sister.

Again, we came to this verse because of the middle letters, the letters that act as the axes of symmetry, in this phrase from the beginning of the Book of Leviticus.

Let us just for a moment give a numerical allusion here, which is very beautiful, in and of itself. The value of the entire phrase "A man who sacrifices from you a sacrifice to God [*Havayah*]" (אָדָם כִּי יַקְרִיב מִכֶּם קָרְבָּן לַי־הוה) is 905. Subtracting the symmetry axes letters that spell "its way" (דַּרְכָּהּ) and whose numerical value is 229, we find that the numerical value of the remaining letters is 676, which is 26 squared. 26 is of course the value of *Havayah*, God's essential Name.

Since this word "her path" (דַּרְכָּהּ) appears here in a very interesting way, when meditating and analyzing this phrase, we are motivated to find where else the Torah relates to this verse and how that helps us better understand its meaning. There is a simple principle that there is nothing that is not alluded to in the Torah. So again, the word "her path" (again, in the feminine form) appears only once in the entire Bible. As we have already said, we find it at the climax of the chapter of wisdom in Job.

God, the uncertainty principle, and Einstein

Now, let us look at this verse, "God understood her path, and He knows her location"[80] (אֱ־לֹהִים הֵבִין דַּרְכָּהּ וְהוּא יָדַע אֶת מְקוֹמָהּ). What is the first thing that comes to mind when relating this verse to our topic of modern science and its relationship to our service of God. This verse is talking about path and location. Quantum mechanics postulates that it is impossible to simultaneously know both the path (which implies the momentum and the velocity) and the position of a particle. This is called the uncertainty principle, the deepest principle of quantum mechanics and the bottom line of this whole theory. In an earlier lecture, we explained that this is not just a problem with the observer's measuring capabilities, this is something inherent in nature. It is simply impossible to know simultaneously both the path (the trajectory) which implies the momentum and velocity vector of a particle and its position.

Now, what is this verse saying? Actually, it is saying exactly what Einstein said when he heard that the uncertainty principle was somehow inherent to nature: "God does not play dice with the universe." It did not sit well with him that God cannot see beyond the uncertainty principle. Little did Einstein know that he had a verse in the Bible to support his intuition that God does know. This is the climactic verse of the chapter of wisdom that states exactly what he felt. Indeed, it says that God does know, indeed God is the only one who knows both the "path" and the "place," i.e., the location.

Now, in this verse, God is referred to by the Name *Elokim*, which we know refers to nature, because it is the Name used in the account of creation appearing in Genesis. *Elokim* refers to the Divinity that is inherent within nature. And what this verse says is that *Elokim*, and only *Elokim*, understands the path and knows the position of wisdom (every elementary particle of nature is a point of Divine wisdom) simultaneously.

What causes the uncertainty principle?

Let us say something further about the implications of *Elokim* knowing the path and position simultaneously. Before the primordial sin, Adam was likened to *Elokim*. Mankind was destined to reach the level of consciousness of *Elokim*, as the verse in Psalms states: "I said [i.e., I desired] that you [mankind] be *Elokim*."[81] But, because you ate from the Tree of Knowledge of Good and Evil prematurely, you fell to the level of just being a mortal man. Had you waited a mere three hours until the beginning of *Shabbat Kodesh*, then it would have been permissible for you to eat from the Tree of Knowledge. That is what it says in Kabbalah. It would have been the "*oneg* Shabbat" (the special pleasure of Shabbat) to eat from the Tree of Knowledge then. (Even though now in the *Shulchan Aruch* it says that there is a special *mitzvah* to taste the Shabbat dishes before Shabbat, when it came to the Tree of Knowledge, God explicitly forbade this.) But because man (Adam and Eve) ate from the tree prematurely, it resulted in the fall of mankind. The continuation of the same verse from Psalms, which refers to this fall says: "Yet, like a man you shall be mortal." Meaning, again that God had in mind that we should attain the level of *Elokim*, but because of the sin we became mortal. This is referring to the human condition.

Now, Heisenberg said that the inability to know the path and velocity of an elementary particle is not just a problem with the observer, with the "human condition," if you will, but it is a limitation within nature itself. In Kabbalah, we learn that when Adam fell from his original state, not only did his human psyche fall, but all of reality, all of nature, collapsed and descended 14 degrees. It is as if there was a wave collapse of all reality when Adam sinned, and this introduced a mortal consciousness into nature, one that does not allow path and position to be known together. Again, originally, before the primordial sin, nature did not limit the observation of a particle's path and position simultaneously.

Now it is even more interesting that this verse is not just telling us that there exists a potential state of consciousness where one can know both path and position simultaneously, it is also telling us how to reach that state. As we shall explain in a moment, symmetry plays a key role in achieving this state of consciousness. If one is perfectly symmetric, one indeed becomes like *Elokim* (as God had initially intended).

Verses with perfect symmetry

To see this we first have to look at the Masoretic note on this verse, a note that was left to us by the final editors of the Bible who lived in the Land of Israel some 1200 years ago, in the time of the *Ge'onim*. They are known as *ba'alei hemesorah*, the masters of the Masoretic tradition. They finalized the Bible in regard to every word's vocalization (*nikud*, in Hebrew) and cantillation marks (*trope*, in Yiddish, or *ta'amim*, in Hebrew). These sages also wrote notes regarding the text of the Bible. These notes are mostly about the frequency of rare words in the Bible. But, on a few occasions they alert us to some very unique phenomena in the Bible, phenomena that no one would ever think of pointing out.

One such example, perhaps the most beautiful, is found on this verse. The Masoretic sages noted that there is a phenomenon in this verse that is repeated three times in the Bible, once in each part of the Bible: once in the Five Books of Moses, once in the eight books of the Prophets, and once in the eleven books of the Writings. What is the special phenomenon? It is that these three verses comprise 7 words each and all three are symmetric around the same middle word, "and he" (וְהוּא). So the structure of these three verses is "$word_1$ $word_2$ $word_3$ וְהוּא $word_5$ $word_6$ $word_7$." What these three verses possess is perfect symmetry. The verse that we have been looking at is the last, the one from the Writings.

The verse from the Prophets is: "A generous person advises generosity and he will stand on his generosity,"[82] or in Hebrew:

וְנָדִיב נְדִיבוֹת יָעָץ וְהוּא עַל נְדִיבוֹת יָקוּם

The generous person is the description of a person who donated to the construction of the Tabernacle. This verse contains another rare phenomenon: the root for "generosity" (נדב) appears three times. A person who is generous himself gives others the advice that they too should be generous. The root for the word "advice" (יעץ) is related to the word for "tree" (עֵץ).

Now, the Mashiach is called one who gives wondrous advice (פֶּלֶא יוֹעֵץ). People are always looking for counseling. Today some counselors are called coaches. The purpose of a good counselor is to rectify your tree. The word for "counsel" or "advice" in Hebrew comes from the word for "tree." The most wondrous counsel, the Messianic counsel, is to be generous. The only one who can give this type of advice is a person who is essentially altruistic and generous. Every person has to have a counselor today, which in Chassidut is called a *mashpia*. Every person today should himself strive to become a *mashpia*. To be a counselor you have to be a generous soul. It is best to have both physical and spiritual affluence from which to be generous. In any case, in virtue of his generosity and his advice to others that they too be generous, the generous soul will stand up.

"Standing up" here also alludes to the resurrection of the dead (תְּחִיַּת הַמֵּתִים). Rashi says that his "stand" is his ability to "stand up" in the national sense, the entire Jewish people standing up, as in the phrase, "I [God] will lead you erect"[83] (וָאוֹלֵךְ אֶתְכֶם קוֹמְמִיּוּת). Interestingly, in Modern Hebrew, the word for "universe" is יְקוּם (*yekum*), like the last word in this phrase. In Modern Hebrew "world" is translated as עוֹלָם (*olam*), whereas "universe" is translated as this word, יְקוּם. Because *yekum* comes from the word קַיָּם, which means "exists," so "universe" is taken to mean "everything that exists." When Adam sinned, he fell, and he will stand up to eternal life in virtue of generosity. So this is the verse from the Prophets.

The verse from the Five Books of Moses that exhibits this symmetry is: "From Asher, his bread is fat, and he will give the delicacies of the king," or, in Hebrew:

מֵאָשֵׁר שְׁמֵנָה לַחְמוֹ וְהוּא יִתֵּן מַעֲדַנֵּי מֶלֶךְ

This is the verse with which Jacob blessed his son Asher. The Lubavitcher Rebbe was very fond of this verse because he was born on the day the prince of the tribe of Asher, (which means "happiness"), brought his communal sacrifice to the Tabernacle—the 11th day of Nisan. The Rebbe explained that the name Asher specifically relates to "commitment with happiness," meaning that accepting the yoke of heaven leads to the greatest happiness.

Jacob, who spoke this verse, is also known as "Israel" (יִשְׂרָאֵל) whose value is 541, and so is the value of first word, "From Asher" (מֵאָשֵׁר). Thus, the name Asher is seen to reflect the entire Jewish people. The verse's second word, "is fat" (שְׁמֵנָה), has the same letters as the word for "soul" (נְשָׁמָה). So these are the three perfectly symmetric Shabbat-like verses with the word "and he" (וְהוּא) in the middle.

Symmetry hidden in the verses

Now let us make a very beautiful numerical observation. The symmetry axis word is "and he" (וְהוּא), and its *gematria* equals 18. But, what happens if we calculate the value of this word's four letters using what is called *mispar keedmee* (מִסְפָּר קִדְמִי), which literally means the "primordial numbering." In number theory, this is what is called a "triangular number." The nth triangular number is the sum of integers from 1 to n. To calculate the triangular value of an entire word we sum the triangular values of each of its letters.[84]

In this case: the first letter *vav* (ו) equals 6, and the triangle of 6 [written: △6] is equal to 21. The next letter is *hei* (ה). It equals 5 and the triangle of 5 [△5] is 15. The third letter is *vav* again and its triangular value is once again 21. The fourth and final letter is *alef* (א),

its value is 1 and the triangle of 1 [$\triangle 1$] is just 1. Finally, the sum of all four triangular values is: $21 + 15 + 21 + 1 = 58$.

As we saw in a previous lecture, the Hebrew word for "symmetry" is חֵן, and its numerical value is 58. So the numerical value of the axis of symmetry of these three uniquely symmetrical verses in the Bible is the same as the numerical value for the word for "symmetry," in Hebrew! We have here a beautiful allusion, and self-reference, to the symmetry inherent in these three verses. In more depth, we might contemplate how these three verses correspond to the three general types of symmetry that we talked about in our previous lectures.

One of the special Names of God discussed in Kabbalah is the Name of 72, which is actually made up of 72 triplets of letters (216 letters in all). One of these triplets is אום. You might note that these three letters are the initial letters of our three verses. So this meditation is an expansion of that Name. So we can now say that this special Name is an allusion to the three different types of symmetry.

Path mentality, position mentality

The Divinity within nature does simultaneously know the position and path of an elementary particle. This is the rectified state of the Tree of Knowledge. What does it mean that there is a Tree of Knowledge of Good and Evil. Why did Adam lust after this tree? The literal understanding is that if you eat from this tree you can know good and evil, simultaneously. As we will explain, to know good and evil simultaneously opens up a third type of dimension that science is not yet aware of. If science would know about this dimension it would most probably solve many, if not all, of the outstanding problems in science.

But for our purposes here, what is the difference between the path that something is taking (a particle, or the path of wisdom) and its position? Path implies velocity and momentum, and place implies position. Above, we talked about the difference between messenger consciousness and matter consciousness; about whether you are a fermion or a boson. Now we will apply this to something similar. There is a person whose consciousness in his life focuses on a path. He is always on the go towards some objective. He is a "walker." A Jewish soul is supposed to be a "go-er," one who is always moving towards a certain goal. This is one of the reasons why *halachah* is called by this name, because just as much as the law seems to be a fixed statute, there is always some inherent flexibility within the law because a law is not static. So this is one type of mentality that people have, being on the go and moving towards a goal.

There is another different type of mentality connected with being static, with having a space. A person wants to have his space, and he is happy in his space. The Rebbe Rashab explained that if a person wants no one else to enter his space, this type of mentality becomes baseless hatred (שִׂנְאַת חִנָּם) and is the reason for the destruction of the Temple and for our remaining in exile for so long.

So here we have a simple way of understanding why these two, movement and space do not go together. When a person is moving, really on the go, he does not ask himself where he is. Either you're on the go or you're stationary. This is a very simple way of understanding the uncertainty principle. To think about where I am requires a feeling of self. When a person is on the move he cannot ask himself, "Where am I?" So you cannot really be moving and ask yourself at the same time, "Where am I?" Only God can have both attitudes at the same time.

These two attitudes are actually the same as "good" and "evil." Good is a good path. To be "good" is to be on the path, to be going somewhere. If you are really on the way, you are really not in any particular place. As soon as you are "someplace," you have stopped. To continually be on the move is the "good news" of the one side of the uncertainty principle. If you are on the go, you do not know where you're at. So not having a place is actually a good thing.

But, if you are concerned about where you are at, you are no longer moving, no longer on the way and do not know where you are going. You have lost awareness of your way. To know your way is good, to know where you are at is bad. So, in this sense, the primordial tree is about knowing "good" and "evil," knowing both where you are at (evil) and still knowing where you are going (good).

The Arizal explains that if Adam would have waited three hours before eating from the Tree of Knowledge, it would have been Shabbat already, and it would have become permissible to eat from the Tree of Knowledge. It would have been a *mitzvah* to eat from this tree. This is because on Shabbat we have Divine consciousness and knowing where you are at (I'm here, but not yet where I should be) actually gives you additional momentum to follow your path in life. In this state, the evil becomes a seat, a foundation for good.

Adding dimensions

Let us now discuss the fact that the three symmetrical verses we mentioned earlier each contains 7 words. The first thing that comes to mind when contemplating the number seven in the context of modern physics and all our considerations of time and space is that most of mankind has set the week for some seemingly arbitrary reason as 7 days. Arbitrary, because nobody actually imagines that nature possesses some kind of phenomenon that corresponds to this cycle of seven days that make up each week. A year corresponds to some phenomenon in the natural world (the revolution of the earth once around the sun), so it is not arbitrary. But, going one step further, even a year is arbitrarily man-made, because it is a cycle that has meaning only on Earth. Outside the Earth, if you were living on Jupiter for instance, one revolution around the sun takes a much longer period of time. So in short, we can generalize and say that all our time cycles, days, weeks, months, and years are not considered by scientists to be inherent cycles in the dimension of time. Instead, they record or measure some local phenomena, such as the movement of the earth, etc.

Above, we noted that when string theorists today add dimensions in order to unify the forces in nature, they only add space dimensions. Nobody today entertains the notion of adding time dimensions. What science knows for sure is that in our world there are three space dimensions and one time dimension.

Now, why do scientists not add time dimensions, only spatial ones? One scientific text even questions this supposition, and indeed asks, why not add time dimensions? The simple answer is that to try to imagine another time dimension is very counter-intuitive—much more so than imagining additional spatial dimensions. Since we already have three space dimensions, it does not seem so strange to add a few more!

Curled-up space dimensions

In order to explain why we do not experience more than three space dimensions, string theorists explain that the additional dimensions are curled up. There is a simple analogy given in books on string theory in order to help us imagine what a curled up dimension might be like. This is called the garden hose analogy. Imagine a garden hose that is strung out between two buildings. If we look at it from a great distance, it seems like a one-dimensional line. But, from the perspective of an ant walking on the hose, there is of course another circular dimension on which it can walk. Now what happens if the ant decides to walk around the hose? From my perspective, from afar, the ant would seem to suddenly disappear; it is as if it went out of existence, because I cannot see that the hose has another dimension. This is the conceptual way that curled up dimensions are described. This is a very simple, *balebateshe* (layman's) explanation.

But now let's think of our garden hose as the dimension of time, not of space. What would it mean to have a curled up time dimension? It would mean that there is some inherent cycle that is running around (or inside) the straight coordinate of time. Maybe this cycle is the week, maybe this is the secret of Shabbat.

Additional dimensions in Kabbalah

Interestingly, the Arizal does not speak explicitly of additional spatial dimensions. However, in the later Kabbalistic writings of the Rashash (who lived in the 18th century), he explains that in order to solve some inherent contradictions that come about because of spatial contradictions in the Arizal's writings, you have to assume that there are indeed more than three spatial dimensions. So the Rashash says explicitly that we have to add *spatial* dimensions to the Arizal's Kabbalah in order to understand it. But, what about time dimensions, the Rashash does not say this explicitly, but actually adding time dimensions in the Arizal's system is much more obvious and natural. This is because the Arizal himself explained that there are different cycles of times, which in the higher worlds are represented by totally different *partzufim* (personas) and it is only natural to attribute a separate time dimension to each of these cycles.

Let us stress that in a certain sense, it is much easier to understand that the curled up dimensions required by string theory are time dimensions and not spatial dimensions. Once more, according to string theory, there are 3 extended spatial dimensions that we experience normally, and an additional 6 curled-up dimensions. Exactly what it means that these curled up dimensions are spatial, is very difficult to understand. They are so small that you cannot detect them, which is why the garden hose analogy is introduced.

The most important thing about time is that there are cycles of 7; e.g., the seven days of the week, the seven years of the *Shmitah* cycle, etc. This was all parenthetical to our main topic, and we could have dedicated an entire lecture just to this subject of additional time dimensions curled up. It can mean something very obvious: that there are inherent, not just man-made, cycles of time, the most important of which is the Shabbat, which is how the world was created. So there is something about time that is based on the number 7.

Shabbat and the symmetrical verses

Usually, Shabbat is considered the seventh day. Sometimes it is considered to be the first of the seven days, from whose holiness all the other days of the week are blessed. But, most often in the writings of the Arizal, Shabbat is considered to be the 4th day of the week. How is this?

From Wednesday to Friday we ascend from the World of Action, to the World of Formation, to the World of Creation. Finally, on Shabbat we ascend to the World of Emanation, the highest plane of consciousness (pure Divine consciousness). Then from Sunday to Tuesday, we bring the Divine consciousness of the World of Emanation (Shabbat consciousness) back down into reality. So on Sunday we descend back to the World of Creation, with the blessing received from the Shabbat. On Monday we descend to the World of Formation, again with the blessing of the Shabbat. Finally, on Tuesday, we descend into the World of Action with the blessing of the Shabbat. For which reason, the Arizal explains that a person who is engaged in work in the physical world should, if he can, limit his work to Tuesday and Wednesday. This would thus endow him with an extended weekend, beginning on Thursday and ending on Monday (with the reading of the Torah on the first, middle, and last day of the weekend). Tuesday and Wednesday (the two days that we are in the World of Action) would remain for devoting oneself to work in this world.

Now, we notice that this order of 3 days, followed by the highpoint that is Shabbat, followed by three more days, has the same structure as that underlying the three verses we discussed earlier: three words, then the word "and he" (וְהוּא), and then three more words.

Lecture 7

In this lecture we will continue our study of elementary particles in light of Torah.

There are three types of elementary particles. Two are referred to as real particles, and the third type is called virtual. As we will describe, these three types of particles correspond to the three Worlds of Creation, Formation, and Action. Virtual particles correspond to the World of Creation. The two other types of particles are bosons (force particles) and fermions (matter particles). Fermions—matter particles—clearly correspond to the World of Action, which is where matter exists, and bosons—force particles—most of which have 0 rest mass[85] (except for the particles of the weak interaction, the W and Z particles, which do have mass) clearly correspond to the World of Formation.

Let us summarize this in a table:

type of particle	World
virtual	Creation
bosons	Formation
fermions	Action

Virtual particles and the World of Creation

Virtual particles are not called that because they are unreal. All physicists agree that they exist. It is because they exist for such a short time that it is *virtually* impossible to identify and observe them. There are four types of virtual particles. They can be virtual bosons (virtual force particles) or virtual fermions (virtual matter particles). They appear in pairs, of positive and negative, as if out of nowhere, and immediately annihilate one another.

There are four sources of virtual particles. Three of these sources are a direct result of the uncertainty principle. These three consequences of the uncertainty principle are:

1) any real particle can spontaneously emit any virtual particle,
2) outer space (which is wrongly considered to be a vacuum) can also spontaneously erupt into virtual particles, which immediately annihilate one another,
3) an electromagnetic field gives rise to virtual particles.

The fourth source of virtual particles comes from Feynman diagrams, where virtual particles are exchanged when an interaction takes place. Quantum mechanics posits that when there is an electromagnetic interaction, two virtual photons, which are messengers, make the interaction take place. A real photon is a particle of light. A virtual photon is "created out of nothing." This is why they are related to the World of Creation. A "vacuum," like space, is continuously creating virtual particles. Another reason virtual particles correspond to the World of Creation is that three of the four sources for virtual particles are a direct result of the uncertainty principle, which we have already identified with the counter-intuition of the World of Creation.

Based on the verse, "The concealed is for *Havayah* our God, and the revealed is for us and our children, forever," Kabbalah teaches that the two higher Worlds, the Worlds of Emanation and Creation, are relatively concealed, whereas the two lower Worlds, Formation and Action, are relatively revealed. So, given that virtual particles are relatively hard to detect and observe, we understand them to correspond to the World of Creation, while force and matter particles, which can be detected, therefore correspond to the Worlds of Formation and Action, the revealed dimensions of reality. Only that the World of Formation is composed of forces, whereas matter exists in the World of Action. So we have here our three types of elementary particles.

Virtual particles and consciousness

Let us repeat that the virtual particles can be either virtual force or virtual matter particles. A very interesting thing is that there are some scientists who believe that conscious systems correspond to the level of virtual particles, or, put another way, that conscious thought is a result of virtual elementary particles, which once more are created *ex nihilo*. This is an amazing relationship. This is another reason to relate virtual particles to the World of Creation, which corresponds to the mind. Conscious systems in nature correspond to the World of Creation.

Now, we said before that with the primordial sin, when Adam ate from the Tree of Knowledge, all of these three worlds fell into a state of uncertainty. So that means that the conscious systems also fell 14 levels. Or, as we said before, that the very consciousness inherent within nature became mortal, which means that it became subject to existential uncertainty. Thus, these three types of particles exist in our three lower Worlds.

Three visions of the Divine Chariot

Now let us look at a more detailed chart based upon the previous one:

<table>
<tr><th>prophet</th><th>consciousness in World of</th><th>type of particles</th><th></th><th colspan="3">type of angels</th></tr>
<tr><td>Isaiah</td><td>Creation</td><td>virtual</td><td>consciousness</td><td colspan="3">Seraphim</td></tr>
<tr><td rowspan="4">Ezekiel</td><td rowspan="4">Formation</td><td rowspan="4">force (bosons)</td><td rowspan="5">Real</td><td rowspan="4">Living beings</td><td>strong</td><td>lion</td></tr>
<tr><td>weak</td><td>ox</td></tr>
<tr><td>e/m</td><td>eagle</td></tr>
<tr><td>gravity</td><td>man</td></tr>
<tr><td>Zachariah</td><td>Action</td><td>matter (fermions)</td><td colspan="3">Horses</td></tr>
</table>

To understand the chart above, we have to know that Kabbalah teaches that there were three prophets who in their prophecy experienced the Workings of the Divine Chariot. They were Isaiah, Ezekiel, and Zachariah.

There are two secret subjects in the Torah: the Workings of Creation and the Workings of the Divine Chariot. "Workings of the Divine Chariot" refers to the *purpose* of creation. Three prophets had a vision of the Chariot, the purpose of creation. Each of them saw the Chariot from a different perspective.

The highest vision was Isaiah's, whose consciousness was in the World of Creation. Ezekiel's consciousness was in the World of Formation. Zachariah saw his vision in the World of Action. Because their consciousness was at a different level, each saw a different image of the Divine Chariot. The components of these Chariots are angels, and in each image the angels were of different types. We already mentioned in one of our previous lectures that angels can be understood as elementary particles. In each vision they saw the Chariot driven by different angelic beings. Of course, the most well-known is Ezekiel's vision, appearing in the first chapter of his book.

We will start from the bottom. Zachariah saw the Chariot as a chariot driven by horses (מֶרְכֶּבֶת סוּסִים).[86] To see the purpose of creation at the level of matter is seeing horses. In the world of Action, it is all just one big horse-race. And the matter particles appear as horses on the run.

The most famous vision of the chariot is Ezekiel's vision. He sees what are called in the Bible "holy living beings" (חַיּוֹת הַקֹּדֶשׁ). His chariot is composed of four different kinds of living beings. Those exist in the World of Formation. How amazing is it that there are four forces in our physical world, which must naturally correspond to these four types of angels. The simple meaning of a "living being" is a being that gives life, or in other words, a life-force. These are of course the four force particles of the four forces in nature. The last of these is the living being which resembles a man. This is not the higher man that in Ezekiel's vision sits on the throne being carried on the chariot. This living being is different from the four others. Living being (חַיָּה) also means "animal," in Hebrew (i.e., animation). So of these four living beings, three are represented by animals, a lion, an ox, and an eagle. So, this fourth one, resembling a man, is obviously different. In the text of Ezekiel, the living being that resembles a man appears first, but in Kabbalah it is ordered last. This living being that resembles a man corresponds to gravity, as we will see. The three others represent the three other forces of nature, which can be almost unified even without string theory. But, the one that is elusive and cannot as yet be connected with the others is gravity. In any case, we have here a beautiful allusion to the four forces in nature.

Finally, the highest vision of the Divine Chariot was experienced by Isaiah (chapter 6), which corresponds to the consciousness of the World of Creation.[87] In the Chariot as seen by Isaiah, there is only a pair of fiery angels (*seraphim*). As we noted, virtual particles appear in pairs. But, what is even more significant for correctly identifying which particles these angels correspond to is the word used to describe them: *seraph*. *Seraph* literally means "fiery angel," which describes them as beings that appear and are then spontaneously burnt up. In other words, they appear and are immediately annihilated, which is the exact nature of virtual particles! Isaiah's vision is of fiery angels that are created and then spontaneously burn away. To burn away, in the case of the *seraphim*, means that they return to their origin in Emanation.

Take the image that we gave above of the vacuum constantly erupting into virtual particles, which are then annihilated. This is an image of something being created (*ex nihilo*, like in the World of Creation) and subsequently returning back to its source. The very same process describes the creation of the *seraphim* out of the World of Emanation, which is the very nothingness that creation *ex nihilo* comes from.[88]

Three types of angels

So now we have three names of angels:

- the *seraph*: the virtual particles in the World of Creation
- the living being, or animal: the real force particles in the World of Formation
- the horse: corresponds to the real matter particles of the World of Action.

Now let us look at the significance of the numerical values, the *gematrias* of these names:

- The value of *seraph* (שָׂרָף) is 580, or 10 times 58, and we are all fully aware by now that 58 is the value of the Hebrew word for "symmetry" (חֵן), so the word *seraph* in Hebrew is a multiple of "symmetry."

- The value of "living being" (חַיָּה), or *chayah* as it is pronounced in Hebrew, is 23. *Chayah* is a very important word in Kabbalah. In the entire first account of creation there are 469 words. The middle word, the 235th word is *chayah*. It appears in the verse, "God said, 'Let the waters swarm with living beings'" (וַיֹּאמֶר אֱלֹקִים יִשְׁרְצוּ הַמַּיִם שֶׁרֶץ נֶפֶשׁ חַיָּה). This is also the first appearance of this word in the Torah. So this word acts as the axis of symmetry for the entire story of creation. This alludes to the fact that while in Ezekiel's vision there is no mention of fish, only land animals, in the Torah the first example of living beings specifically refers to fish.
- Finally, the value of "horse" (סוּס) is 126. The letters of this word are themselves symmetric. There is something very symmetric about a horse. (Apparently a horse can go back in time, retaining CPT symmetry).

So we have now seen that each of these three words exhibits a different form of symmetry.

The sum of all three types of angels, *seraph* (שָׂרָף), living-one (חַיָּה), and horse (סוּס) is 729, which is also 27 squared, or 9 to the third power, or 3 to the sixth power. Whenever we have a number raised to a power, an expression in the form x^y, we are expressing a quality of inter-inclusion.

The Paschal lamb, matzah, and bitter herbs

We are taught in Kabbalah that there is another set of three very connected words whose sum is 729. In the *Hagadah* we quote from the Talmud that whomever does not experience three things on Passover has not performed the *seder* properly. These three words are: the "Paschal lamb" (פֶּסַח) whose value is 148, *matzah* (מַצָּה), whose value is 135, and "bitter herbs" (מָרוֹר), whose value is 446.

Their sum, again, is 148 + 135 + 446 = 729! Obviously, this is a meditation to have on Passover.[89]

Let us see how these three components of Passover correspond to the three types of particles. Indeed, today the Paschal lamb remains at the virtual level for us, because we cannot offer the sacrifice. The *mitzvah* of the Paschal lamb is that it be eaten roasted, more burnt than anything else eaten in the Temple service. There is no other offering that is eaten so "burnt," so *seraph*-like. This represents the level of consciousness attained by a *tzadik* who always feels how the world is coming into being out of nothing and instantaneously returning to nothingness.

Eating the *matzah* corresponds to the four force particles. Bread is what gives a person energy, or life-force. As the verse states: "Not on bread alone can man live, but on the word of God will man live."[90] So it is not the physical matter that gives life, but the inner life-force. *Matzah*, which is a form of poor bread, reveals this—that it is not the physical but the spiritual within the physical that gives life. So eating *matzah* is to remind us that real *force*, real life, comes from the utterance of the Almighty.

The matter particles correspond to the bitter herbs, and remind us of the hardship that is inherent in the world of Action. Hardship is ultimately a good thing, because reward comes from the hardship that we experience. In the verse that lists the four Worlds, the World of Action is preceded by the word "even" (אַף), which also means "anger," alluding to the difficulty and hardship inherent in the physical world. So once again, we see here a very nice correspondence of the three elements of Pesach with the three types of particles, with the three lower Worlds as the foundational frame of reference for all these correspondences.

Unifying the particles in each world

We mentioned in an earlier lecture that there are three methods for unification:

- in the World of Creation, the unifying principle is nullification (הִתְבַּטְּלוּת), we see this as the *seraphim* being burnt, that is nullified, in their return to nothingness.
- in the World of Formation, the unifying principle is inter-inclusion (הִתְכַּלְּלוּת).
- in the World of Action, the unifying principle is cooperation (הִשְׁתַּתְּפוּת).[91]

Corresponding these three methods with the three visions of the Chariot, we may now say that the horses (i.e., the fermions, the matter particles) of the World of Action need to cooperate, to act, or work together. The unification of matter is when it works together.

Unifying the four forces of nature

Let us now turn to the four forces in nature. As we have just said, all four are included within the World of Formation. The living beings (i.e., the bosons, the force particles) of the World of Formation are joined through inter-inclusion. In Ezekiel's vision, each of the living beings exhibits perfect inter-inclusion, because each has four faces, with each one of the faces resembling one of the four living beings: a lion, an ox, an eagle, and a man. And furthermore, each of these faces inter-includes all four in itself (just that the predominant face is one of the four). Thus, the unifying principle for force particles is inter-inclusion.

The paradigmatic enigma today in physics is how to unify the four forces of nature. We mentioned in a previous lecture that this is what physics is still lacking: a model of unification based on inter-inclusion. This type of unification is readily apparent from a holographic image that is etched on a crystal. In other words, somehow in the strong force you should be able to see the weak force, the electromagnetic force and gravity, all reflected in it.

The *Targum Yonatan*, based on the sages says that together the four *chayot*, the living beings, had 256 wings. 256 is 4 to the fourth power (4^4). This is a beautiful example of inter-inclusion, because, as mentioned, each of the four living beings had four sides to it, and each side had four faces, and each face had four wings. So altogether the four living beings had 4 to the fourth wings.

There is a particular type of symmetry known as angular symmetry, if you look at something from a different angle you will still see the same thing (like in a ball under rotation). Here, from whatever direction you looked at the living being, you saw the same thing. So in that sense, wings are like spin.

The *seraphim*, the fiery angels of Isaiah's vision, each had 6 wings. The horse angels of Zachariah's vision each had 2 wings. They are the most like how we usually image that angels have 2 wings (like physical birds).

Now we can draw the following correspondence:

particle	# of "wings"
virtual	6
force	4
matter	2

This correspondence and its scientific and spiritual implications need to be developed further. But, let us once more note that the living beings exhibit the greatest degree of inter-inclusion, where all four living beings reflect all four and each of these reflections itself reflects all four. And altogether they have 256 wings.[92]

Time reversal in Ezekiel: the dry bones

The book of Ezekiel,[93] begins with a vision that is related to the number 4 (four faces, four wings, etc. to each *chayah*, each living being). Close to its end there is another vision related to the number 4, known as the vision of the dry bones.[94] The dry bones represent the dried up state of the Jewish people while we are in exile. In his vision, Ezekiel sees the entire House of Israel as dry bones, which means that they resemble the dead. Not only has death set in, but the remaining bones are scattered apart, representing a state of maximum entropy where everything has fallen apart completely.

Then Ezekiel describes something which is an example of time-reversal: the dry bones come together and sinews, flesh, and skin grow on them. When scientists want to demonstrate the concept of the arrow of time—that time has a direction, from past to future—they tell us to imagine a shattering glass. Now if this shattering was filmed and played backwards, we could immediately tell that the film was playing in reverse as we would see the pieces coming together. This is an example of *teshuvah*.

It is also similar to the principle that in the Torah, in the Pentateuch, there is no chronological order,[95] in the normal sense of chronology. Sometimes, things that happened later are written as if they happened earlier. This concept is the foundation of *teshuvah*, which allows a person to go backwards and relive, and even take positive advantage of negative past events. In this manner, as already explained, evil becomes a seat for the good.

Ezekiel sees the film of "death" being played in reverse, the bones coming together and growing a human being as it were. This is the experience of resurrection. Ezekiel's vision is an example of time reversal and a shattering of the law of entropy. From a state of high entropy (a later state, as death progresses) a state of lower entropy emerges (an earlier state, the energy of life). The arrow of time normally travels from low entropy to higher entropy. And here, in Ezekiel's vision he sees the opposite occurring: a high entropy state (death)—a state with less order—develops into a low entropy state (resurrection), a state with more order.

In his vision, Ezekiel saw the body divided into four components: bones, sinews (blood vessels), flesh, and skin. As he saw the parts being added to the bones, entropy was decreasing, but at the end the body was still not alive—life-force was still missing. So if these four parts of the body represent the four forces of nature, then something more still had to be added in order to unify the forces together, to bring the body to life. This is what Ezekiel prophecies as: "From all four corners, the wind [spirit] shall come to enliven these dead corpses."

Now, we cannot go into great detail for lack of time, but let us see how these four parts of the body correspond to the four forces:

- The bones correspond to the strong nuclear force, which holds the nucleus of the atom together, just as the bones (the skeleton) hold the entire body together.
- Blood vessels (sinews) correspond to the weak nuclear force. It is the weak force that is responsible for radioactivity, which bears an important relationship to blood, that we will not discuss right now.
- Flesh, or the muscular system, corresponds to the electromagnetic force. In principle, the nervous system corresponds most closely to the electromagnetic force, and it is most closely represented in Ezekiel's model by the muscular system.

- Skin corresponds to gravity. The complexion of the skin represents the curvature of space. As we explained, according to Einstein, space is relatively gently curved and its curvature is equivalent to gravity. We mentioned that the biggest discrepancy between general relativity and quantum mechanics is that relativity speaks of gentle curvature while quantum mechanics speaks of frantic curvature. Einstein was into the World of Rectification, which is gentle, while Heisenberg was into the World of Chaos, which is frantic. In any event, it is a very beautiful image to see the skin as representing the gentle curvature of space. The skin is where you see the human features of the face (just as in the four faces of the living beings—the angels in Ezekiel's vision of the Divine Chariot—the one that resembles a man, corresponds to gravity).

The fact that then the spirit has to come in to unify these four forces, to make them into a living man, this is similar to the creation of man, where after Adam's body had been formed physically, God breathed into him the breath of life,[96] which then animated him and made him alive.

Lecture 8

Wave-particle duality

The first topic that we will discuss in this lecture is the wave-particle duality of elementary particles. Einstein was the first to postulate that light, which at that time was considered to be a wave phenomenon, possessed a particle expression called a photon. Unlike an electron, a photon has no mass. It travels at the maximal speed, the speed of light (c) which for Einstein in his theory of relativity is considered to be the absolutely fastest speed physically possible in this world.

Later, another scientist discovered that particles, like electrons, also exhibit wave properties. In a certain sense, this was a greater innovation, but since Einstein had already said that a wave is also a particle, this is usually seen as just the other side of a single duality. This is an example of symmetry (in this case a duality). If what seems to behave like a wave also exhibits particle-like behavior, it is simply the next step to say that what seems to behave like a particle also has wave-like behavior.

The wave function is a probability function, which measures the probability that a particle be found in a particular position (or state) in the universe. Now it is agreed that all elementary particles, whether they are fermions or bosons, exhibit this duality, which is a paradoxical phenomenon. It only depends on whether the observer is concentrating on observing the wave or the particle-like characteristics.

Let us explain what this duality means in terms of Chassidic teachings. What does it mean to be a wave and a particle at the same time?

One important model in Kabbalah, which is an example of the principle of "a reversing seal" (חוֹתָם הַמִּתְהַפֵּךְ), the reversing seal (a principle of symmetry) which applies to every Kabbalistic model that makes use of the essential Name of God, *Havayah*. Whether it be the four letters as they correspond to wisdom, understanding, the emotive faculties, and kingdom, or whether it be the model of how they correspond to the four Worlds, the model always exhibits this property. Between them, the four letters of God's essential Name exhibit the phenomenon of a reversing seal, where whatever corresponds to the first two letters, is duplicated in reverse in relation to the two last letters. The best example to illustrate this is how the four letters of God's essential Name correspond to love and fear:

Yud – higher awe

Hei – higher love

Vav – lower love

Hei – lower awe

The metaphor for this is that when you have an embossed seal (the stamp sticks out) then when you push down on hot wax for instance, it will engrave an inverted image of itself.

In respect to elementary particles, let us first say that we can relate their particle and wave properties to two concepts that appear clearly in Kabbalah: matter and form. That the particle properties correspond to matter is clear, and here we are adding that the wave properties correspond to form. Now, the example here of how the reversing seal is used in a model relies on recognizing that in our revealed reality form *informs* matter and therefore is considered higher than matter. Revealed reality corresponds to the two lower/last letters of the Tetragrammaton, *vav* (ו) and *hei* (ה). In almost all discussion of these two concepts we find that form is like the soul of matter. So in our revealed reality, form is higher than matter. Form is like the male aspect and matter is the female aspect. But, in respect to their origins, the reverse is true: the origin of matter is higher than the origin of form.[97] The origins correspond to the two first/higher letters, *yud* (י) and *hei* (ה) of God's essential Name, *Havayah*. These two levels are also called the father and mother principles. So the origin of matter, which is called by the sages *hiyuli*[98] matter, or primordial, ethereal matter, is like the father and the hidden origin of form is like the mother.[99]

So our complete model, which exhibits the reversing seal, looks like this:

Letter of Tetragrammaton	
yud (י)	(origin of) matter
hei (ה)	(origin of) form
vav (ו)	form
hei (ה)	matter

Furthermore, we are taught that the subjective experience of matter arouses awe and fear, whereas the experience of form arouses love and attraction. Meaning, if we experience light as a wave, in some way, it arouses love. This is exactly what is said that in the two higher levels, the *yud* is called the higher fear, which is nullification of reality connecting it with the primordial matter, whereas, the *hei* (mother) is called higher love, which corresponds to the origin of form.

But in the revealed levels, form is the masculine, while matter is considered the feminine. So in direct experience, the masculine induces love, while the feminine awakens fear, or awe.

The sages say that innately a child fears and has awe of his father and loves his mother. But this needs to be balanced, so that he should also fear his mother and love his father. To this end, when the Torah commands us to fear our parents, it first mentions the mother: "A man should fear his mother and father." But, when it commands us to honor our parents, and honor is considered to be an expression of love, then it first mentions the father: "Honor your father and mother…." Once more we see that the father represents awe and the mother represents love (these correspond to the two initial letters of the Tetragrammaton). But, in respect to the husband and wife relationship (corresponding to the two last letters of the Tetragrammaton), the husband becomes love and the wife represents awe (as in Proverbs, where the good wife is called a God-fearing woman[100]—because she assumes the simple duty of raising her children in accordance with the Torah). So, this is a very general principle throughout the entire Torah. The phrase most often used to describe this principle of the inverted seal is in Aramaic and it reads: "fear and love, love and fear."[101] Often, before commandments that we perform we say, or have in mind, that we are performing this commandment "with fear and love, with love and fear." This is just like saying: "matter and form, form and matter," or "particle and wave, wave and particle."

According to Kabbalah if there is a true state of unification, it must have both manifestations, both aspects. If it is the higher unification, then the particle comes before the wave. If it is the lower unification, then the wave is above the particle. But, each unification has to have both aspects to it. Especially, if it is a total consummate unification, it has to have all four aspects to it. Meaning that this would be an objective of science to try and distinguish between the concealed wave property, vs. the revealed wave property. And even harder: to distinguish between the concealed particle property vs. the revealed particle manifestation. The clue to understanding these distinctions is that the concealed property is "non-local" whereas the revealed property is "local." And therefore, the concealed wave/particle property envisions each wave/particle as relating to and interacting with all others across the span of the entire universe simultaneously, while the revealed property does not. Locality vs. non-locality is a most important topic that we leave for later.

Einstein in the Bible?

In this context, let us mention that James Gleick, the author of *Chaos*, claims that the three most brilliant physicists of all time lived in the previous century. They were all Jewish and they all suffered from anti-Semitism in one way or another. They were, Albert Einstein, Lev Landau, and Richard Feynman.

Lev Landau was a Russian physicist. In addition to being a physicist he was an educator. Gleick mentions that when Lev Landau taught physics he created a system to rank the genius of a physicist. It was like a quantum system and went in increments of ½. The greater the physicist, the lower his number. Landau gave Planck, Bohr, and Dirac a ranking of 1 (as he did for Richard Feynman). He ranked himself initially as 2.5, but at the end of his life he ranked himself as a 2. According to Landau, in all of history, Einstein was the only physicist worthy of a ½ rank. This conceptualization of rank for physicists goes very well with what we discussed in an earlier lecture, that the smaller you are, the greater you are.

Apropos, let us say another thing. If Einstein is at a different level altogether than all the other physicists, so he too must be alluded to in the Torah. What we are trying to say is that everything is alluded to in the Torah. Nowadays, people look for celebrities and events in the Torah using equidistant letter skipping and the like. All those so-called Torah codes, which there is a lot to say about, but not right now, are not as significant or as serious as finding some explicit source in the Torah: a literal source in the Torah for a person or a concept. For Einstein there is a very explicit and beautiful such source.

In the previous lecture, we discussed Zachariah and his vision of the Divine Chariot. There is another important prophecy by Zachariah who lived in the time of the second Temple's construction, after the Babylonian exile. In his prophecy, he sees the cornerstone of the Temple and the Menorah. About the cornerstone he writes: "On one stone are seven eyes"[102] (עַל אֶבֶן אַחַת שִׁבְעָה עֵינָיִם). This one stone that he is referring to is the cornerstone (or headstone) of the Temple. The end of the prophecy reads: "The headstone, grace, grace to it"[103] (הָאֶבֶן הָרֹאשָׁה תְּשֻׁאוֹת חֵן חֵן לָהּ). The word translated here as "grace" is the Hebrew word for "symmetry" (חֵן). Since here it says "grace, grace," this even means double symmetry. Now, the name "Einstein," in German and Yiddish literally means "one stone." This idiom actually appears five times in the Bible. If we would like to give a lecture on Einstein we would have to go through all five times that the Bible speaks of "one stone"[104] (אֶבֶן אַחַת). But, now we will only mention the first and the last.

The first time is found in the Book of Joshua where it mentions that when the Jewish people were crossing the Jordan river, entering the land of Israel, each of the princes of the 12 tribes took one stone on their shoulder from the riverbed and carried it to the Jordan's west bank.

The last time that this phrase "one stone" appears in the Bible is here in this prophecy of Zachariah. What does it mean here that the "one stone" has "seven eyes." There are some commentaries who say that the seven eyes are a symbol for the guarding force watching over the "one stone." But, in our opinion there is a much better explanation.

In another place in the Bible, we find that the phrase "seven eyes" symbolizes God's Providence over every single aspect of creation (what we know as personal and individual Providence). So, what the verse is saying is that all of God's Providence over the entire world is concentrated through the cornerstone of the Temple. It is like the "one stone" is the focusing lens of Divine Providence.

This concept also appears in two other verses in the Torah. The first is a verse in Deuteronomy, which states that "the eyes of God are upon the land of Israel, from the beginning of the year to the end of the year". So here, the land of Israel is like the lens through which God sees the entire world. First His "gaze" is focused on the land of Israel, and from there it spreads out to the rest of the world.[105] The second verse is recited as part of the ceremony when bringing of the first fruit to the Temple: "Gaze down from Your holy abode in the heavens and bless Your people, Israel." In that case, it is explained that "the heavens" are a metaphor for the Torah, thus the Torah acts like a lens through which all of Divine Providence passes.

In any case, what this means is that God's Providence is first concentrated on the "one stone" of the Temple, and then spreads out to every minute detail of creation.

Seven space-time directions

But now, what do "seven eyes" have to do with Einstein. We mentioned earlier that Einstein united space and time. Normally, we consider space and time to be 4 dimensional: 3 dimensions of space and one dimension of time. But, the time dimension is not the same as the three space dimensions. What is the difference between them? When represented mathematically, as a complex field, the time dimension is described as the imaginary component (in terms of i, the square-root of negative 1, or $\sqrt{-1}$) in this complex plane, while the spatial dimensions make up the real component. But, more simply put, the difference between the time dimension and the space dimensions is obviously that in each of the spatial dimensions you can go in two directions—to two different extremities. I can go east and I can go west. I can go up and I can go down. There is no preference or dictation in which direction to go. But time is uni-directional. I cannot go in both directions. This is the arrow of time, which we have explained already as the law of entropy.

Another way to understand direction is as a vector, which is like the direction of sight. So each of the seven eyes is symbolic of sight in a given direction, like a vantage point from which something is being looked at. So in 4-dimensional space-time you can actually only go in seven directions.

As we saw in the previous lectures, in Kabbalah we always speak of the 6 extremities of space, not just 3 spatial dimensions. But, time only has one direction. These seven directions correspond to the seven attributes of the heart and to the seven days of the week. The Midrash tells us that each day of the week is paired with one another, just like the three spatial dimensions have two directions each. But, Shabbat complained to God that it was lacking a partner (exhibiting its uni-directional character). So, the Midrash says that the Almighty paired the Shabbat with the Jewish people. The six days, which correspond to the six extremities of space relative to Shabbat, are paired. But, Shabbat, which is the essence of time, does not have a pair, so that is why God paired it with the Jewish people.

What does this mean? It means that there is something about the congregation of Israel that can go backwards in time. Because if they are really going to be the Shabbat's soul mate so, even though

Shabbat is uni-directional, nonetheless, the Jewish people represent movement in the opposite time direction. How is this? Because the Jewish people exhibit the secret of *teshuvah*, they can go back in time. Indeed, Kabbalah stresses that the word Shabbat in Hebrew can also read as the word that means "[you shall] return" (שַׁבְתָּ), the literal meaning of *teshuvah*, to return. *Teshuvah* is capable of converting a transgression into a merit. But to do that you have to be able to go back in time. So *teshuvah* is like a time machine.

Shabbat: time travel and teleportation

In Hebrew, "time machine" is written מְכוֹנַת זְמַן, and has the numerical value of 613! This is of course the number of commandments in the Torah. If a person is complete in all 613 commandments then he has a time machine. There is a related concept called teleportation, which is the short form of tele-transportation. This is a phenomenon of quantum tunneling. The way to think of a teleporter is that it does with space what a time machine does with time, so a teleporter is actually a "space machine." In Hebrew we would call a "space machine" מְכוֹנַת מָקוֹם, and its gematria is 702, the value of "Shabbat" (שַׁבָּת).

So here we have an example of the inter-inclusion of space and time within Shabbat. On the one hand the Shabbat complained to God that it has no partner, indicating that it is uni-directional, or time-like. On the other hand, we see here that Shabbat equals a space-machine, meaning that it is space-like.

This reminds us that the sages say that if the Jewish people would keep but two Shabbatot, they would immediately be redeemed. Sometimes it is explained that the two Shabbatot are the two Shabbatot before Pesach, *Shabbat Parashat Hachodesh* and *Shabbat Hagadol*, because Pesach is the time of redemption. But, in Kabbalah and Chassidut it is explained that these two Shabbatot occur together on the same Shabbat, meaning that every Shabbat has two aspects to it. This corresponds to our description of Shabbat having a potential of moving us freely both forward and backward in time. Shabbat is the seventh eye on the one stone that represents the uni-directional arrow of time. But, in order to be redeemed we have to pair it up with the Jewish people through *teshuvah*, specifically *teshuvah* out of love. Redemption is freedom. The ultimate prison that the world is in right now is time. The best you can do is to maximize your time. But, to get out of the uni-directional prison of time, to free yourself from aging, that is the ultimate form of redemption. Only the Jewish people can achieve this freedom and redemption through the power of *teshuvah*. Because the Jewish soul is the Shabbat's partner, it is like the second extreme of the dimension of time.

One stone and seven eyes

So all of this came out of our searching in the Bible for an allusion to Einstein. The "one stone" is the lens through which all the personal Divine Providence goes through. We have 7 eyes peering or gazing through the one stone. The model here is 1:7 and the word that reflects this ratio is "then" (אָז), whose two letters, *alef* (א) and *zayin* (ז) equal 1 and 7, respectively. Now what is the sum of one "stone" (אֶבֶן) and 7 times "eye" (עַיִן)?

אֶבֶן עַיִן עַיִן עַיִן עַיִן עַיִן עַיִן עַיִן = 963

But, 963 is also the value of the phrase found in the beginning of creation: "And God saw that the light was good"[106] (וַיַּרְא אֱ־לֹהִים אֶת הָאוֹר כִּי טוֹב).

Recall that "time machine" equals 613. The first time there are adjacent words in the Torah whose numerical value is 613 is in this very phrase, specifically, "[And God saw] the light" (אֶת הָאוֹר). So, light is a "time machine," as above. All this was in praise of Einstein. We hope that he appreciates it.

How do you say "wave-particle duality" in Hebrew?

Our motivation for going into this discussion about the allusion to Einstein in the Torah is that every concept in science must have an origin that can be identified in the Torah. One might think that a sophisticated idea like light being both a wave and a particle simultaneously would be beyond the Torah's mind-set, that there could be no simple word to describe this. But, *chas veshalom* that this should be the case. This is a false thought. There must be a word in the Torah that expresses the wave-particle duality inherent in a photon. Why are we so sure of this? Because we are taught that God creates the world continuously through the letters of the holy language of the Torah, so if there is such a phenomenon it must be possible to find it in the Torah.

One of the foundations of teaching science in a 100% Torah context is that we need to rename many of the concepts and terminology that science uses today.

We now are looking for a word in the Bible that unites the concepts of wave and particle. And we would like this word to be directly related to light, which is the way that Einstein originally came upon this duality. So we turn to the Book of Job once more. Job is a man afflicted with terrible suffering. He has three very good friends who come to console him, but fail. Then a young man, who was listening quietly to the conversation, interjects. His name is Elihu ben Brachel (אֱלִיהוּא בֶּן בְּרַכְאֵל), whose numerical value is 358, the same as "Mashiach" (מָשִׁיחַ). After Elihu speaks, God himself addresses Job with 50 rhetorical questions dealing with the secrets of creation: "Do you know this?" "Do you know that?" God's message to Job is that if you do not know the secrets of creation, you cannot understand the Divine Providence that governs your fate.

In the introduction to his commentary on *Sefer Yetzirah*, the Ra'avad explains that these 50 questions correspond to the Fifty Gates of Understanding. One of those 50 questions is: "Who gave birth to the drops of dew?"[107] (מִי הוֹלִיד אֶגְלֵי טָל). This is one of the most beautiful of these 50 questions. Drops of dew are very special because they do not fall from the sky like rain or snow. Instead, we wake up every morning to find these drops on the grass. Where did they come from? From our perception they came out of nothing. Earlier, we talked about virtual particles that just appear out of nowhere. Our experience of drops of dew is that they just come out of nothing. The word we are going to be focusing on is "drop" (אֵגֶל).

Every word in Hebrew has both a three-letter root, and an essential root, called a gate, which is its two-letter sub-root. Now, the first experience that we have of light is as a wave. "Wave" in Hebrew is a familiar word, גַּל (pronounced "*gal*"). This is indeed the grammatical gate, or two-letter root, of the word "drop" (אֵגֶל). How so? There is a rule in Hebrew grammar that there are seven letters (האמנתי"ו) that do not change the essence of the essential two-letter root; they are the vowels (אהו"י) and the three letters מנ"ת, which tend to fall from a three-letter root leaving the two-letter gate, or essential sub-root.[108] So definitely the letter *alef* (א) that precedes the two letters *gimel* (ג) and *lamed* (ל) is an added letter to the two-letter sub-root גל. Some of the commentaries explain that drops of dew are related to waves, because when you look at an entire field full of grass and all the drops of dew on it, it looks like waves of dew.

Now, if you look in the Radak's book of Hebrew roots, you will find that אגל is a root with only a single meaning, a drop of dew. It is never used in any other context. It is not a drop of rain, or a drop of oil, or a drop of anything else; only a drop of dew. What is special about this is that the sub-root by itself means "wave", but as soon as you add an א to it, it becomes a "drop" (אֱגֶל).

A drop of dew is special because it just seems to appear out of nowhere. But more importantly, dew is a symbol in the Bible for light, as in the verse, "Your dew is the dew of light"[109] (טַל אוֹרֹת טַלֶּךָ). The reference in this verse is not just to regular dew or light, but to the dew, the light with which God resurrects the dead. This is a very special type of light called "Torah dew" (טַל תּוֹרָה), alluding to the hidden mysteries of the Torah, the deepest secrets of the Torah, that bring the dead back to life.[110] Thus, God is asking Job, do you know who gave birth to these drops of dew? We have here a most beautiful origin for the fact that photons have a dual property, a dual representation as a wave (גָּל), and as a particle (אֱגֶל). For this reason, we can say that the best choice for a word for "photon" in the language of the Bible, is אֱגֶל, pronounced *eigel*.

This word should not be confused with the Hebrew word for "calf" (עֵגֶל), which is pronounced similarly but is written with an *ayin* (ע), not with an *alef* (א).[111]

The value of changing scientific nomenclature

Let's return to a story we did not tell before about Richard Feynman. As a child he did not speak until he was three years old. Sometimes the greatest geniuses start speaking at a very late age, so do not worry if your child talks late. Sometime in grade school, Feynman taught himself trigonometry. He saw that trigonometry uses all these strange notations like sin(e), cos(ine), tan(gent), etc, which take three letters to write. At a very early age he came up with a much better notation. He decided to use his completely different and simpler notation. Much later in life he came up with diagrams to describe some of the most complex ideas in quantum mechanics. But, at this early age he was very frustrated that nobody adopted his notation, which he considered to be much better. At a certain point he matured enough to understand that he cannot change the world. Even though I could have done it much better, it is a lost cause.

So the question we have to ask is: should we too despair of changing the word used by all of science today to fit the Torah? We don't think we should despair! One of our objectives in teaching science is to change words. It's a very difficult objective. Feynman tried and failed. If it is already engraved in the psyche of the world it is difficult to change it.

But, here we are saying something else. We are looking for the correct Torah terminology for describing something that science has discovered. So we want to change "photon" into "*eigel*," and "electron" into "*ofan*,"[112] and there are numerous other such examples. Now, the moment that science adopts the Torah word for these terms it itself will be resurrected. Scientific thinking will be given new life, like the dry bones we were talking about earlier. To resurrect science, it has to use the proper Torah terminology.

More waves

Now, the two-letter sub-root of the Hebrew word for photon (אֵגֶל), *eigel*, that means wave is as we said, גַּל, pronounced *gal*. We explained that there are seven letters that you can add to the sub-root that do not significantly change its essential meaning. For example, we mentioned the word that means "calf" and is written with an *ayin* (ע) added to it. But, since *ayin* is not one of these seven letters, it significantly changes the essence of the sub-root. In our case, there is another letter, except for *alef* (א) that most easily fits in phonetically to this sub-root: the letter *mem* (מ). The *mem* can be added either before the two-letter root, in between the two letters, or after them. In each case, we have a new word that has meaning. They are

- Adding the mem before yields the word מַגַּל, which means "sickle"; if we have our high grass with the drops of dew on the top, then we take a sickle, it is a symbol for many things in the Torah.
- If you put the *mem* in the middle of the wave it turns into גָּמָל, which means "camel."
- And if you put it at the end of the wave, you get גֹּלֶם, which means "golem."

So all three of these words are variations of the two-letter root meaning "wave" (גל) with a letter *mem*.

In the next lecture we will see how camels are themselves symbols of photons.

Lecture 9

The camel as a metaphor for wave-particle duality

We all remember the story of Isaac and Rebecca and how Abraham sent Eliezer, his servant, as a messenger (like we learnt before, regarding messenger particles) to find a wife for Isaac. The example brought in the *mishnah* we discussed earlier in respect to sanctifying a wife by messenger is in this story, which very often is explained in Kabbalah and Chassidut as the most essential *shiduch* (match) in the entire Torah. Everything that we want to learn about *shiduchim*, about matchmaking comes from this story. When Eliezer had made the *shiduch* with Rebecca, even her father and brother, who were not very righteous individuals, had to admit that "from *Havayah*, this thing came" (מֵי־הוה יָצָא הַדָּבָר). In Hebrew, the value of this phrase is 378, which is also the value of the word חַשְׁמַל, a word that in modern Hebrew has been adopted for "electricity."

Parenthetically, if you reverse the numerals of 378, you get 873, which is the numerical value of the idiom "the speed of light" (מְהִירוּת הָאוֹר), the one essential constant of our physical reality according to modern physics. A similar example is that if you take the value of "Shabbat" (שַׁבָּת), 702, and write the numerals in reverse order, you get 207, which is the numerical value of "light" (אוֹר).

Every word in the Torah has a particular place where it is especially concentrated, meaning, that the reality that this word represents is concentrated maximally in this place in the Torah. Surprisingly, the most ubiquitous word in the verses describing the *shiduch* of Isaac and Rebecca is "camel" (גָּמָל), another of the images for a photon. A camel is both a particle and a wave, because if you look at its hunch, it is like a wave. But if you take in the whole camel at once, it itself is a big particle. We mentioned that it has been shown that all elementary particles exhibit wave-particle duality, so in essence they are all camels.

Meditation on the meanings of "camel"

All phenomena of the Hebrew language are amazing meditations. The same three-letter root that means "camel," גמל, also means "weaning." In the Torah, weaning appears specifically regarding Isaac. The Torah tells us that on the day that Isaac was weaned Abraham made a celebration[113] (בְּיוֹם הִגָּמֵל אֶת יִצְחָק), another connection between Isaac and this root, and hence another connection between Isaac and wave-particle duality.

Besides meaning "camel" and "weaning," this same root (גמל) can also mean "acts of loving-kindness" (גְּמִילוּת חֲסָדִים), something that is specifically related to Abraham, Isaac's father. So what might be the meditation linking the three meanings of this root? The vast majority of the times that this root appears in the Torah is in the sense of "camel." Incidentally, the English word for camel clearly comes from the Hebrew word *gamal*. There are certain words in the English language which are a hundred percent from Hebrew.

$E=mc^2$

The same three-letter root, גמל, is also simply the name of the third letter in Hebrew, the *gimel* (ג). The English alphabet in a certain sense simply mirrors the Hebrew alphabet. In English the third letter is c, which is the first letter of "camel."

What about in physics? In physics c denotes the speed of light, as in $E = mc^2$. It is easy to understand that E stands for "energy" and m stands for "mass," but what does c stand for? Even physicists are divided on the reason for the usage of c in this convention. One opinion says that c stands for "constant." Another opinion says that it stands for the first letter for "speed" in Latin.

[But we know what it really stands for: c stands for camel! So, $E = mc^2$ stands for energy is equal to the mass times a squared camel! It's hard to square a camel because he's round, but this is an important thing for women to know. A woman who wants to have a calm husband can use a technique brought by Rabbi Avraham Abulafia, which involves rearranging the letters of a word or phrase to get something related. In this case, the letters of "calm husband" (בַּעַל רָגוּעַ) can be rearranged to spell "circle the square" (עִגֵּל רִבּוּעַ). A woman who wants to calm her husband down has to know how to make a square into a circle. This is not just cute. In Chassidut, it is explained in length that people are uptight and tense because they have too many corners, too many sharp edges. To calm them down, you have to help them round their edges and get them to change from a square into a circle.]

In the Pentateuch, there are 34 instances of the root גמל. They do not always mean "camel." As we have noted, sometimes, they mean "wean" or "generous." The significance of 34 for us is that the Hebrew word for "photon," *egel* (אֶגֶל) is equal to 34. The value of this word is twice the value of "good" (טוֹב), reminding us of the verse: "God saw the light that it was good."[114]

Of all those 34 times, 18 are concentrated in the story of Abraham sending Eliezer with the camels, with gifts loaded on the camels.[115] On average, almost every third verse mentions camels. And Eliezer, the messenger particle, says that he will test the woman to see if she is worthy of marrying the son of the most generous soul on earth, i.e., Abraham. Will she give my camels water to drink? And that was the way that Rebecca proved herself worthy of entering Abraham's household, by watering the camels. 18 is the numerical value of "live" (חָי). So this story is all about the "live camel."

Sight and camels

At the end of the story, Eliezer takes Rebecca on the camels back to the land of Israel, where Isaac is waiting. The Torah relates that Isaac had gone to *Be'er lachai ro'ee* (בְּאֵר לַחַי רֹאִי). Upon his return home, he went out one evening to pray in the field, he lifted his eyes and saw Eliezer and his camels arriving. The sages explain that while Eliezer had gone to find Isaac his wife, Isaac had gone to this place called *Be'er lachai ro'ee* where Hagar, Abraham's banished wife was living, in order to make a *shiduch* for his father, Abraham.

Be'er lachai ro'ee literally means "the well of the Living One that was seen," but it can also mean the "well of living sight," an allusion to photons of light coming out to give life (as in photosynthesis). We can therefore say that Isaac seeing Eliezer's camels coming over the desert is like an image of photons, camels, arriving. Isaac is like a scientist who is watching the particle and wave effect. He was contemplating sight, and light, and the electromagnetic radiation, something that is hinted to in the place that he was coming from: "the well of sight...."

When Rebecca saw Isaac and how beautiful he was, the Torah says that she fell off the camel. It is like she fell off of her photon. She was "riding" the beam of light and fell off of it. There are many explanations of what it means that she fell off the camel. In order to get married, Isaac had to lift his eyes and see the camels coming.

Now let us return to the word "camel" itself. We saw that it is formed by adding the letter *mem* to the Hebrew word for wave (גָל). The addition of a *mem* to the wave gives the wave a particle reality, just as the addition of an *alef* to גל, forms the Hebrew name for photon, and gives the photon's wave reality, a particle reality. Now, why is the addition of a *mem* very good? Because, the meaning of the letter *mem* (when treated as a whole word) is "water" (מַיִם)! Clearly, the first image of waves that we have from the natural world is from the waves of water.

The sickle as a metaphor for wave-particle duality

We saw in the previous lecture that there are other ways to create a three-letter root out of the two-letter root, גל. In "camel" (גָּמָל), the *mem* is added between the two letters גל. What about the addition of the *mem* before these two letters, which then spells "sickle" (מַגָּל)? The form of the sickle also resembles a wave. Moreover, the sickle is used in the field, which relates it to another important element of our topic: fields. Today, science is looking for a unified field theory.

When searching for a bride for Isaac, Eliezer tested Rebecca's loving-kindness, her ability to do kindness with the camels (as mentioned, גמל also means גְּמִילוּת חֲסָדִים, acts of loving-kindness).

We saw another root that comes from adding the *mem* to the end of גל, "golem" (גֹּלֶם). A golem literally refers to a substance without form. It can also mean an unformed mind. In Pirkei Avot,[116] the Ethics of the Fathers, there is a description of the golem as a potentially wise man, who in reality is not a stupid person, but rather one who has not finished his training yet, his mind is as yet unformed. Because he is on his way, everything for him is counter-intuitive. But once he finishes he will become a wise man.

Indeed, what is important about these three words—"camel" (גָּמָל), "sickle" (מַגָּל), and golem (גֹּלֶם)—is that they share the same numerical value, 73, which is also the numerical value of "wisdom" (חָכְמָה). Wisdom is the origin of light. In Chassidic teachings it is called "the beginning of revelation" (רֵאשִׁית הַגִּלּוּי). The numerical value of the first verse of the Torah is 2701, which is the triangle of

73, the triangle of "wisdom" (חָכְמָה).[117] So the first verse of the Torah is like a giant triangular camel. The value of "camel" is thus very important. We should keep in mind that so far this is the largest image of wave-particle duality.

The particle zoo

Camels are force particles, while the horses that we discussed earlier are matter particles. To recap, we said that the three types of particles (real matter, real force, and virtual particles) correspond to horses, living beings (animals), and fiery angels (*seraphim*), respectively. So now we are adding that the force particles are likened to camels, which act as messengers. We are translating all the particles into a zoo.[118]

What about the virtual particles? What animal should they be likened to? Answering this question requires that we expand our imagination. To do so, we need to have already rectified our power of imagination, an important subject in and of itself in Chassidic teachings.

We explained that there are both virtual matter particles and virtual force particles. In the World of Creation there are both virtual Formation (force) particles and virtual Action (matter) particles. So, actually we need to find two different animals as metaphors for both types.

Without explaining this too deeply, the virtual matter particles would be likened to "elephants" (פִּילִים). The Talmud[119] says that if you dream of elephants, you should expect wonders of wonders to happen to you (wonders are actually greater than miracles). The most wondrous experience is to see an elephant pass through the eye of a needle (פִּילָא בְּקוּפָא דְּמַחֲטָא). This reminds us of the double slit experiment, which ascertained the wave-matter duality of real photons. Here we are talking about virtual particles.

We saw that according to string theory, size does not matter, so there is no difference between imagining photons passing through a microscopic slit, and imagining elephants passing through the eye of a needle. The elephant is considered to be the biggest animal. In the Song of Creation (which relates the song that every part of creation sings in praise of the Creator—the song reveals the essence of each part of creation), it is written that the elephant sings the verse: "How big are your actions, God,"[120] alluding to itself as the biggest of God's creations. So a virtual electron, which is a virtual matter particle, would be likened to an elephant. This is what comes to mind in regard to the virtual matter particles.

Now, what about the virtual force particles, like a virtual photon? These we say are symbolized by a butterfly. In Hebrew a butterfly is a *parpar* (פַּרְפַּר). Virtual particles come in pairs, and the word *parpar* is like "pair pair." They also spontaneously appear and disappear, which is also "appear disappear." These are two linguistic puns on the Hebrew word *parpar*. In any event, now we have all this imagery of horses, camels, elephants, and butterflies in our quantum zoo.

Camels in the Bible

We mentioned already that in the Five Books of Moses there are 34 instances of the root גמל. In the entire Bible there are 112 instances of "camel." Now, since the numerical value of גמל is 73, looking at the 73rd instance of this root in the Bible is most significant (as the 73rd instance of a root that equals 73 is an example of self-reference). The 73rd instance is in the verse: "God is God of payments, he will surely repay"[121] (כִּי אֵ־ל גְּמֻלוֹת י־הוה שַׁלֵּם יְשַׁלֵּם).

There is a famous saying of the sages in the Talmud, "'Knowledge' is great, for it appears between two Names [of God]." There is a verse: "God is God of knowledge."[122] In the original Hebrew, the word "two types of knowledge" (דֵּעוֹת) appears between two

Names of God, *Kel* and *Havayah*. The sages then mention that the word "Temple" also appears between two different Names, *Adni* and *Havayah*, which they also explain. The Talmud then goes on to ask about the word "revenge." Since it too appears between the same two Names as "knowledge," it too should be great. As we have discussed in the past, the teaching about "knowledge" can indeed be extended to "revenge." Very significantly, when one checks the entire Bible, one finds that there is one more example of a word that appears between two Names of the Almighty. This is our word here: "payments" (גְּמֻלוֹת). Surprisingly, though the sages do not mention this.

Now this word, "payments," even though it stems from the same root, does not mean "camel" here. It means "payment," a meaning that is closer to the meaning of "acts of kindness." But payment can be positive or negative. You can get a reward and you can get punishment as payment. Though it does not mean "camel," "payment" is still etymologically related to camel, since they stem from the same root. It is in the *Zohar* that we find that the camel can be a negative image. Sometimes it is so negative that it is a symbol for the angel of death (that should be written on Camel cigarettes). Nonetheless, camels were part of the positive image of the patriarchs and matriarchs, especially Isaac and Rebecca. The camel is also a symbol of acts of loving-kindness. The unified view of a camel is therefore that it is a symbol of just payment: reward or punishment.

The essential message of the letter *gimel* (ג) is that of reward and punishment, as explained in length in our book *The Hebrew Letters*. To believe in reward and punishment, the eleventh of the thirteen principles of faith enumerated by Maimonides, is camel-consciousness. That is what the camel represents: reward and punishment. Punishment is also reward, because it too has a positive intent to it—it purifies. This is what this phrase mean: "...God is a God of payments."

So, now we have that the three words that appear in a verse between the two names *Kel* and *Havayah* (this excludes "Temple," which appears between *Adni* and *Havayah*) correspond to the three axes of left, right, and middle.

Right (loving-kindness): "God is God of payments."

Middle (knowledge): "God is God of knowledge."

Left (might): "God is God of revenge."

The simplest reason explaining why the sages do not note our phrase is because it is so similar to the verse on "revenge," except that revenge is pure might while payments alludes to loving-kindness as well. So now we have the basic understanding that all the forces of nature are camels.

Space as a desert

To complete our meditation on this image of the camel as a force, let us ask what medium the camel travels through? The camel is known as the "ship of the desert." Its medium of travel is the desert. For a camel the space-time continuum is the desert. So the desert represents the medium through which the forces travel, which is space-time. Thus the camel, which represents the photon, the force particles, traverses the desert, which represents space-time. Just like space, the desert has curvature. The desert is in a sense full of waves. We might ask: Why do camels have humps? Because they are born of the desert! If you are born out of something, you resemble it, you represent it. The camel, which is born out of the curvature of the desert, resembles its origin in that it has a hump.

Another example of this principle that we have already seen is the drop of dew. We said that it seemingly comes out of the nothingness. So the drop itself resembles nothingness. Seeing a drop of dew is like seeing (as much as can be seen) the nothingness from where it originates.

The Dirac Sea

The famous physicist Dirac, symbolized space-time as a sea—even called the Dirac Sea. He called the medium in which the universe exists: a sea. A sea is also full of waves.

What is the connection between the desert and the sea? Based on Psalm 107, the sages learn that four individuals should thank God for delivering them from a dangerous situation.[123] One is a person who has successfully made a journey through the desert and another one who has a made a journey by sea. Dirac envisioned the medium of the universe as a sea. Likewise, one can imagine the entire world as a gigantic desert, part of which we inhabit. Just as a sea has islands in it, so a desert has its own islands: oases. The universe also has islands, which are stars, planets, moons, etc.

Space as a field

There is another concept that is very similar and actually appears more commonly in science: the image of a field. Every force establishes a field. A field also has a wave-like property to it. Like grass growing in a field and then you have to take the sickle to cut down the grass.

How do you say "space" in Hebrew?

You can imagine space-time in modern physics symbolized as a desert, or a sea, or a field. There is one more word that is used by the sages to connote "outer space," in Hebrew. That word is חָלָל (*chalal*). The first step of creation is the contraction of the infinite light which results in the creation of "the empty space" (חֲלַל הַפָּנוּי) within which the entire universe was created. Let us give all these images a sign, or acronym.[124] The particular sign in this case is made up of the initial letters of these four symbols for space-time: "desert" (מִדְבָּר), "field" (שָׂדֶה), "sea" (יָם), "space" (חָלָל), which spell the word Mashiach (מָשִׁיחַ). *Mashiach* has now become our image for the four symbols of space-time in Torah.

Space is not empty!

Let us return to Dirac's sea. What was he thinking of when he chose this symbol. He was imagining fish jumping up and down. The fish symbolize virtual particles that are continuously being created and annihilated. According to quantum mechanics there is no such thing as a vacuum, because everywhere there are eruptions of virtual particles, which are so-named because they are short-lived, not because they do not exist. So every little piece of space-time is constantly creating and annihilating virtual particles. This is a sea of particles. This is similar to the way the Torah describes the creation of fish and reptiles from water, from the sea, "Let the waters teem with crawling, living creatures...."[125]

The Hebrew word for "space," *chalal*, is usually translated as "vacuum." But, quantum mechanics says that there is no real vacuum. What does Torah have to say about the existence of a vacuum?

Before her complete conversion to Judaism, Rachav, the innkeeper, said to the spies that she knew that, "your God is the God of the heavens above and the earth below."[126] When Moshe Rabbeinu stated a similar concept he said it with the exact same words, but added another two at end: "there is no other" (אֵין עוֹד).[127] The sages explain that what Rachav did not know is that there is no real vacuum—because the Almighty is even in the vacuum, or in Hebrew "even in reality's vacuum" (אֲפִילוּ בַּחֲלָלוֹ שֶׁל עוֹלָם).[128] There is no other, even in the apparent vacuum of the universe. Meaning, that God is equally present even in what seems to be a vacuum. In essence, there is no real vacuum.

Rachav did not have this understanding until she became a complete convert. This also teaches us about the difference between a full-fledged convert and a non-complete convert. The full-fledged convert receives a Jewish soul in his or her internal consciousness and therefore can understand this concept.

The word used for "vacuum" in Hebrew, *chalal*, is also the root of the word "event," as in מִתְחוֹלֵל, some activity is going on. So it is very significant that in Hebrew, the language of creation, the very word for "vacuum" implies that activity is going on; something is happening in the vacuum. This is exactly the mindset of quantum mechanics. Space looks at first as if nothing is there, but there is an infinite amount of activity going on inside. This is a beautiful example of insight into nature using the Hebrew language itself.

So we have four words in Hebrew that can be used to describe the space-time medium. Because the acronym designating these four words spells "Mashiach," we may conclude that our becoming conscious of the nature of the medium of the universe is an indication of having Messianic consciousness.

According to many modern physicists, space-time itself is in essence gravitons, it is gravity itself (gravitons are the as yet undiscovered force particles of gravity). From these gravitons erupt all of the other particles, spontaneously, as virtual particles do.

Lecture 10

Classic Determinism

Determinism is one of the great questions faced by modern science. If we believe that we can perfectly know all the causes and their effects, then we should be able to know everything. If we could just put all the initial data into a man-made computer, we should be able to predict all future events. This understanding of determinism, which indeed contradicts the concept of free-will–the basis of the Torah–was a dogma of classical science. Laplace, one of the greatest scientists of the previous century, wrote clearly that if you could tell him the exact position and movement of every single particle in the universe, then he could tell you everything that would happen in the future.

This definitely contradicts the entire Torah, the entire faith system that is built on the principle that man has free choice, and that man is free to determine his own path in life—that it is not predetermined, that it is not fate. If there is a reason to argue that science contradicts faith, it is not the age of the universe, or some other external thing that will eventually be explained by science itself. It is only because of this problem, the problem of determinism. Evolution is not *the* problem, unless it is entirely based on determinism.

Quantum determinism

Classic determinism was the dogma of science until quantum mechanics. So quantum mechanics was a very good innovation. Quantum mechanics declares that it is impossible to really know the initial conditions, all the more so that I cannot know where things will end up. So, initially it seemed that quantum mechanics was the downfall of classic determinism. About which, we may say: "*Baruch Hashem*"; "*Shehechiyanu*." We can make a blessing that quantum mechanics came along and destroyed classic determinism, for our good, for the good of the Torah, for the good of faith.

But, then quantum mechanics came up with another form of determinism, which is not the same; it seems to be more o.k. than the classic determinism. This is called quantum determinism. What is this? Quantum determinism says that though I cannot know exactly what will happen, I can still know pretty much (maybe 99%) what will happen, because of probability considerations. The foundation of quantum mechanics is probability. Probability is the essence of the wave function, which is actually a probability function. So if there is a tremendous probability that something will happen, I can say with almost complete certainty that it will happen, although there is a miniscule chance for freedom of choice.

This is also sometimes called probabilistic determinism—everything is probable, not 100% determined as thought before quantum mechanics. Even Einstein had a problem with this little bit of choice that was left. Einstein was a Jew, so he should have preferred this quantum determinism, but he had a real problem with it and he preferred classical determinism. It was very hard for him to accept that God plays dice with the universe. We explained previously that there is an inherent mortal consciousness in nature that is the result of Adam's primordial sin and the descent of nature into an uncertain state.

In any case, this is the second stage of determinism: though science no longer believes in classical determinism, it accepts quantum determinism.

Black holes and determinism

But, then along came Stephen Hawking, who is the greatest authority on black holes. Black holes were already predicted by general relativity, so Einstein knew about them more or less. A black hole is a region of space that has so much matter and so much gravitational pull that even light that approaches the black hole will be drawn and sucked and swallowed by the gravitational field surrounding the black hole. That is why it is called a black hole, because light cannot reflect off of its surface, it is dark. It is the ultimate prison, whomever enters cannot leave. Everything has a verse to describe it. In this case the verse that describes such an inescapable prison is: "All who enter will not return."[129] This verse actually refers to heresy. Whoever comes into it will not be able to get out of it. That's what Egypt appeared to be. No slave could escape. It was the absolute prison. Nonetheless, the Almighty delivered us out of Egypt.

Hawking's first insight was that because of Dirac's sea (which we discussed in the previous lecture), the continuous activity in the apparent vacuum of space, there forms what is called Hawking radiation or glow. Remember that because of quantum uncertainty, pairs of virtual particles are constantly being created spontaneously everywhere in space. Since these pairs of virtual particles comprise a particle and its anti-particle, normally, they just annihilate each other immediately. But, in the case that they are created exactly on the event horizon of a black hole—the event horizon is the theoretical barrier beyond which anything entering will not be able to escape the black hole's gravitational field—then one particle is ejected and one is sucked into the black hole's gravitational field. Since one is ejected, even though the black hole cannot be seen, it does have a distinctive glow around it, made up of these ejected virtual particles. This theory is of course all mathematical. No one has actually observed a black hole. This was Hawking's first insight into black holes. This is very interesting. You would have thought that anything that is black could not be seen. The color black is such because it absorbs all colors of light, yet somehow, here, black does glow, something comes out of it.

As we will explain, black holes are related to how the entire universe was created, and it is surmised that every galaxy has a black hole at its center, including our own galaxy.

Then Hawking had a second insight, which has arguably fueled one of the greatest disputes in the scientific world today. It has to do with information and with the question of if, when a virtual particle is sucked into a black hole, its wave function is forever lost and unrecoverable. This idea is controversial, because if it is correct then quantum determinism does not hold. If I would know the state of the universe now, which is all the information, then I could by probability predict 99% correctly the future. But, if a chunk of information is lost, it would destroy the possibility of determining the future, even within probability. This is because quantum determinism assumes that all of the information (all of the possible outcomes of the wave function) is still present. But, if all of a sudden the universe loses information, due to the black hole, then even probability cannot tell me about the future.

String theory on determinism

Perhaps the greatest innovation of string theory—even greater than the notion of strings, or more dimensions, etc.—has to do with this paradox. String theory says that even after wave destruction, the information lost in a black hole can be retrieved. Not only does a black hole cause radiation to be emitted around it, but even the information that is swallowed up in it can reemerge, meaning that the information is not lost forever. This would seem to return us to quantum determinism.

So if retrieving the information is like resurrecting it, Hawking believes that once something is buried in a black hole, it can never be resurrected. But, string theory believes, like the Torah, in resurrection.

Returning lost information

So all this has tremendous scientific ramifications. But, our goal here is to answer how all of this is alluded to in the Torah, and how it stands in relation to the Torah.

There are two simple and obvious images for "lost information." The first is that losing information is like forgetting. To lose information psychologically is like forgetting. So the question about information lost in a black hole is: is there something inherent in the universe that allows it to forget. According to the Torah, it is very good to forget, because by forgetting you can open up new directions of choice. But, if the world is never going to forget all of its previous information, it will be a much more deterministic world.

The second, even more straightforward image for information loss is of losing, like misplacing something (in Hebrew, אֲבֵדָה). As King David says in Psalms: "I am lost like a sheep, seek Your servant."[130] Like "lost and found." When a person loses something, there is a custom to say, "*Elaka de'Meir Aneini*" ("God of Me'ir, answer me") and then he will find it. To lose psychologically is to forget; to lose a physical object is just an אֲבֵדָה.

One of the most important commandments in the Torah is that if someone has lost something and you find it, you have to return it to him. This is a commandment governing our relationship with other people. So if the universe also "loses" things (information), we might ask if we are obligated to return them to it.

Lost and forgotten

The initials for "something lost" (אֲבֵדָה) and "something forgotten" (שִׁכְחָה) spell the Hebrew word for "fire" (אֵשׁ). The sages use the image of finding something that has been lost as an image for the whole topic of finding one's soul-mate. Finding your spouse is like finding your back-side (your lost "rib"), as originally Adam and Eve were one person joined back to back. Indeed, in Hebrew, the word for woman (אִשָּׁה) etymologically stems from the word for "forgotten" (נשה). The backside represents the unconscious, the "hole" where things are forgotten. There are seven synonyms for "earth" in the Bible.[131] One of them is נְשִׁיָּה, pronounced "*neshiyah*,"[132] and meaning "the place where things are forgotten." These are idioms that are thousands of years old and which are black-hole like.

Let us look at the numerical values of these two words:

- Something that is lost is אֲבֵדָה, whose numerical value is 12.
- Something forgetten, שִׁכְחָה, is equal to 333.

Their numerical sum is 345, which is the numerical value of the name "Moshe" (מֹשֶׁה).

This is quite strange as Moshe Rabbeinu signifies memory. The verse in Malachi says: "Remember the Torah of My servant Moshe." Moshe is the epitome of mind and wisdom. So if Moshe equals "forgetfulness" and "loss" together then that means that he is their opposite. This is a case of a rectifying *gematria* where one side of the equivalency rectifies the other.[133] So it is very interesting that these two images for information lost in a black hole equal the value of Moshe Rabbeinu's name. This is once again one of the most important issues in all of modern science, as it can change our entire understanding of determinism in nature, which is especially relevant in respect to science in relation to faith.

The black holes darkness and glow

Now, the root of the word for "something forgotten" is: שכח. The most important permutation of this root is "darkness" (חֹשֶׁךְ). This is the literal meaning of a "black hole," eternal darkness, a symbol of death, or being buried forever. But, most significantly, these two words "darkness" and "forget" have the same letters in Hebrew.

The first time that darkness appears in the Torah is even before the creation of light. Before God created light, the earth was dark. This follows the natural order of things as stated by the sages: "first darkness then light."[134] In respect to science, what this means is that first comes the darkness of the black hole and only then, as a secondary phenomenon, does the light, the glow (Hawking radiation) begin to come out of the black hole. This is the type of light that is described in Ecclesiastes as, "light is more beneficial [when it comes] from darkness."[135] The order in creation is first darkness then light.

The event horizon

A black hole's most important phenomenon is its border, its event horizon. What does the Torah say about the primordial darkness? "And darkness on the face of the abyss."[136] If there is a term in the Torah that refers to the event horizon of a black hole, this is it. The "face of the abyss" is the border of the abyss, where the abyss of course is a beautiful metaphor for a black hole. Everything is trapped in there. What is the only thing that can come out of there; that we can resurrect? It is lost information!

In the Talmud, the image for a case of lost information is that of a *talmid chacham*, a scholar, who has forgotten his learning.[137] Even if it seems to have totally disappeared, it is destined to reappear. That is why the sages say that you still have to honor him, even if he seems not to remember anything. It will ultimately be retrieved. It will resurface. This resurfacing of information is alluded to in the final part of the verse describing the "event horizon": "And the spirit of God hovered over the face of the waters." And then "God said: Let there be light...."

So we have here beautiful images in the beginning of the Torah that represent black holes, event horizons, and darkness before light.

So we could say that all of science (vis-à-vis faith) revolves around the question of whether the universe forgets or not, and whether that forgetfulness is permanent. The answer to this question seems to hinge on whether the wave collapse that the virtual particle entering the black hole experiences is complete or not. If it is complete, then essentially, the information is not retrievable. If it is not, the information may reappear elsewhere, or could be retrieved from the black hole itself.

To date, no one knows how to interpret quantum mechanics in respect to the reality that we experience every day. There are many competing interpretations. Two of these are the Copenhagen interpretation and the many-universes interpretations and they differ on the question of the necessity of wave collapse. According to the Copenhagen interpretation, wave collapse is essential to the quantum mechanical take on our reality, because without it, reality would not be strictly defined and things would stay in a state of indeterminacy. But, according to the many-universes interpretation, wave collapse is not essential as *all* possible outcomes occur, each in a different universe.

Let us say that Hawking's notion about black holes swallowing up information forever is found in the verse: "death swallows [things] forever."[138] Whereas string theory's notion of information being retrievable from a black hole corresponds to the verse: "It has swallowed strength and spit it out"[139] (חַיִל בָּלַע וַיְקִיאֶנּוּ).

The point we are trying to emphasize is the concept of forgetting. It says that the Almighty remembers all that has been forgotten. Everything that has been destroyed is still remembered by God.

We said that sometimes it is good to forget, because it opens up possibilities of choice. Sometimes it's good to forget, as the Ba'al Shem Tov says that in general it is not good to know your previous incarnations as it will limit your free-will. The Arizal said that it is good to know your past incarnations in order to focus on what your rectification is in this life time. But, the Ba'al Shem Tov felt that such knowledge may well fixate you on a certain image of what your life should be about. It would limit you to a certain rectification. Consequently, you will not be free to progress infinitely in this lifetime to greater things than just seeking one particular rectification. Death is a classic parallel to information that has been swallowed up by a black hole and the question of retrieving the information from the black hole is like the process of remembering one's previous incarnations.

The paradox of free-will and determinism

Let us now address the paradox of determinism and free-will. It is important to know that in Judaism determinism and free-will are not an either/or proposition. They are paradoxically both true simultaneously. Everything is predetermined and yet, there is free-will. This is because Judaism is inherently paradoxical.

In Jewish philosophy, as for instance in the thought of Maimonides, the two sides of the paradox are known as "[God's] knowledge and [our] choice," or "[God's] decree and [our] choice." But, in the original way that this paradox was stated by the sages, the paradox is not between God's knowledge and our choice. There is only one way that this paradox is phrased, which is not the normative philosophical way of putting the question in an abstract way. This singular phrase, which states the paradox of the most important spiritual problem of life, is: "All is foreseen, and permission is given"[140] (הַכֹּל צָפוּי, וְהָרְשׁוּת נְתוּנָה).

This expression is from Rabbi Akiva, the Moshe Rabbeinu of the Oral Torah. He was the only one who could express the essential nature of this paradox in words. This is very poetic language. The poetry here is very significant. It is not a dry, dead intellectual question. It is saying simultaneously: "all is foreseen" and "permission is given." "Foreseen" does not mean the same thing as "known," i.e., that God knows all. It certainly does not mean "decreed" (גְּזֵרָה), meaning that God has decreed that we act in a certain way.

By the way, the usual answer given to the problem of knowledge and free-will is to distinguish between "knowledge" and "decree." Meaning that decree is much stronger and affects our actions much more than knowledge does, but that in reality God only knows, He does not decree. "Determinism" usually implies "decreed." But, in Rabbi Akiva's statement the word used is neither knowledge nor decree.

Now, whoever has a fine knowledge of Hebrew knows that the translation of this word is not even "foreseen," but rather "foreseeable." This is a nuance in Hebrew. It is like saying that "this was foreseeable," I should have been able to guess this. Now, this sounds a lot like quantum determinism. It is a very probabilistic way of seeing. But, you are given permission to do what you like. Again the word here is not that you have free choice, or free-will, but that you have permission, you have a green light, to act as you like.

We could go on in length about this point. But, now let us make a "sign" for this teaching.[141] The symbol for this paradox is the bird, as the Hebrew word for "bird" (צִפּוֹר) is an acronym for the two key words of this phrase: צָפוּי רְשׁוּת. God sees everything in a foreseeable way, and everything has freedom to act. So now we have a symbol of a bird to go together with our particle zoo.

The *Zohar*[142] says that the Mashiach's soul, before it comes down, resides in the "bird's nest" (קַן צִפּוֹר). The nest itself represents consciousness since the phrase "bird's nest" (קַן צִפּוֹר) has the same numerical value as "consciousness" (מוּדָעוּת). If I would have perfect consciousness, then I would be like a bird in the nest, like the Mashiach, able to exercise complete freedom of will, but he is still in the nest waiting to fly out and come to us. Like a chick that is waiting to mature sufficiently to fly into the world, into our minds and hearts.

The diagonal of an n-dimensional cube

Let us see some slightly more sophisticated *gematria* than what we have until now.

Let us begin with a little understanding of what happens when you add dimensions to space-time. When you add a dimension you are really going up, numerically, by square roots.

One of the basic principles in Kabbalah[143] is called the "twelve diagonals" (יב גְבוּלֵי אֲלַכְסוֹן).[144] If I have a unit square, the length of its diagonal is $\sqrt{2}$.

If it is a square with length and width n, then the diagonal will be $n\sqrt{2}$.

In any case, $\sqrt{2}$ is to be associated with the diagonal.

Rabbi Levi Yitzchak of Berditschev points out that the word "diagonal" (אֲלַכְסוֹן), relates to "seeing the nothing" (סְכֵל אָיִן). So if you can see the "diagonal" of something, you are seeing the aspect of "nothingness" in it. If you want to see nothingness, do not look straight, look diagonally.

All diagonals are multiples of square roots.

If I have a unit cube, then the diagonal is $\sqrt{3}$. To compute this we see that it is the diagonal of the diagonal of two of the sides and one more dimension. Or in other words:

The diagonal of two sides is $\sqrt{2}$, as we said above. And then the diagonal of that with another of the dimensions is, following the Pythagorean theorem:

$$\sqrt{(2+1)} = \sqrt{3}$$

Now, for space-time, which has four dimensions, the diagonal is $\sqrt{4} = 2$.

But, the Book of Formation defines that there are 5 dimensions, 3 space, 1 time, and 1 value dimension which runs from good to evil (we mentioned that if science would take this dimension into account, many of the problems it encounters today would be solved). The diagonal of a unit 5-dimensional cube would be $\sqrt{5}$. This is also the basis of the golden section, as explained in our book on the golden section. There are books that claim that the most important number in the universe is $\sqrt{2}$ and there are those who claim that that number is $\sqrt{5}$.

So first, let us state the general principle: $\sqrt{n}$ is the diagonal of the cube in n-dimensional space.

So the diagonal of an 11 dimensional cube is $\sqrt{11}$. And, again, the diagonal represents the nothingness from which all somethingness comes.

In string theory, which now surmises 11 dimensions, 10 of which are spatial, the $\sqrt{10}$ is also very important. This would be the diagonal of space in string theory.

Square roots, the commandments, and God's essential Name

Now let us write out the three most important square roots to four decimal places:

$\sqrt{2} = 1.414$

$\sqrt{3} = 1.732$

$\sqrt{5} = 2.236$

If we take out the decimal points, as is done in Kabbalah, we get that: $2236 = 2 \cdot 1118$ or 2 times the value of the *Shema*, "Hear O' Israel, *Havayah* is our God, *Havayah* is one" (שְׁמַע יִשְׂרָאֵל י־הוה אֱ־לֹהֵינוּ י־הוה אֶחָד).

$2236 = 86 \cdot 26$, where 86 is the value of God's Name *Elokim* (א־להים) and 26 is the value of God's essential Name, *Havayah* (י־הוה). 1118 is the first number that is a multiple of both 86 and 26. This number 2.236 is the basis of the Golden Ratio, which is the basis of the beauty and elegance of the universe.

Now, the sum of the two other numbers is also a multiple of 26, the numerical value of God's essential Name, *Havayah* (י־הוה): $1414 + 1732 = 121 \cdot 26$, or $11^2 \cdot 26$. It follows then that the sum of all three is "light" (אור) times "God" (י־הוה), or 207 times 26.

What is the geometric relationship between 86 (the value of "*Elokim*") and 207 (the numerical value of "light")? If we draw a square with length 86, the diagonal will be 121, which is the difference between 207 and 86.

Now, let us see what happens when we add $\sqrt{5}$ to $\sqrt{3}$. We get, $2236 + 1732 = 248 \cdot 16 =$ רמח · הוה, where 248 is the number of positive commandments and 16 (הוה) are the final three letters of *Havayah*. Taken as a stand-alone word, their meaning is "present" (for, according to Kabbalah, the positive commandments rectify the present moment of reality, the inherent reality of the three created

Worlds of Creation, Formation, Action, which correspond to these three final letters of *Havayah*). Now, adding $\sqrt{5}$ to $\sqrt{2}$, we get: 2236 ⊥ 1414 = 365 · 10 = י · שסה, where 365 is the number of prohibitive commandments and the letter *yud* (י), whose value is 10, is the first letter of *Havayah*, which corresponds to the World of Emanation (pure Divine consciousness, above experience of created time and space), the level of consciousness reached by our observing the prohibitive commandments (the level of "no" above "yes," while simultaneously giving rise to the "yes").

Lecture 11

Wormholes in tapestries and space

The Talmud in the tractate of Shabbat explains that there are 39 categories of work that are forbidden on Shabbat. These are based on the 39 categories of work that were done in the Tabernacle. The Tabernacle is considered to be a miniature of the world. Science would do well to take a close look at these 39 categories of work and ask what each one corresponds to in nature.

Now, the 24th is tearing in order to re-sew.[145] The Talmud asks: Where do we find this category of work taking place in the Tabernacle? The Talmud replies that this occurred when a worm-hole appeared in one of the Tabernacle's tapestries and then the fabric had to be torn and re-sewn. When you see this passage in the Talmud in relation to any book on modern physics today, it totally blows you away.

Today, modern science describes space-time as a fabric which can theoretically be torn, as if a worm hole, and then re-sewn. And one of the main questions is whether indeed tears in the fabric of space are possible and whether they could be used to create "wormholes," theoretic passages through space that would shorten distances considerably.

There were two layers of tapestries in the Tabernacle. One layer had 10 tapestries in it, and the other had 11 tapestries. In modern string theory there are two types of theories: five theories that posit 10 dimensions and an 11 dimensional model (called M-theory), which unites all five by showing that each is a particular case of this more general model. Indeed, in the Tabernacle too, the lower layer of tapestries had 10 tapestries and the upper had 11.

Leaps of distance

Tosafot, the medieval commentary, notes that the only thing that could happen to the tapestry was a wormhole forming in it and that this was the only place that a wormhole could occur in the Tabernacle. A wormhole in physics is like a "leap of distance" (*kefitzat haderech*) by the Ba'al Shem Tov.

The worm (once again from our particle zoo) is a symbol for King David, who describes himself as a worm: "I am a worm, not a man."[146] The Ba'al Shem Tov was the one who used to travel through "wormholes" on his journeys, and he is considered to be a reincarnation of King David.

In any case, according to general relativity, the fabric of space-time cannot be torn, the topology can change gently, but it cannot be torn. But string theory allows for this. We are discussing this in short, but the entire issue here in the Talmud has to be studied deeply.

Even more amazing is the fact that in the entire Talmud, there is only one place where it says that there is a commandment to study science. We find this statement immediately after the discussion of the tearing of the tapestry. The Talmud says:

> Said Rav: One who knows how to compute *tekufot* and *mazalot* and does not do so, you are not permitted to quote his teachings in his name....
>
> Said Rabbi Shimon son of Pazi: Rabbi Yehoshua ben Levi said in the name of Bar Kapara: One who knows how to compute *tekufot* and *mazalot* and does not, about him the verse writes, "They do not gaze at the acts of God, and His handiwork they do not see."

> Said Rabbi Shmu'el bar Nachmani, said Rabbi Yochanan: From where do we learn that man is commanded to compute *tekufot* and *mazalot*? From the verse, "And you shall keep and perform these, for they are your wisdom and understanding in the eyes of the nations." What is wisdom and understanding in the eyes of the nations? The computing of *tekufot* and *mazalot*!

According to the *Smag* (an acronym for the Medieval index of commandments called: *Large Book of Mitzvot*), this is one of the 613 commandments. In his words: "It is a positive commandment to compute *tekufot* and *mazalot* and *moladot*...."[147] These three words "*tekufot*, *mazalot*, *moladot*" refer to the various cycles of the stars and planets (today, we call this astrophysics).

On the verse, "for it is your wisdom and your understanding in the eyes of the nations,"[148] *Rashi* writes that science is "recognizable wisdom," because all the nations of the world can recognize it.[149] Science has an advantage over the wisdom of the Torah because it predicts. The Torah does not appear to predict anything and only the Jewish mind fully appreciates it. Torah is like concealed wisdom compared to science. As we know, the strength of any scientific theory is its power of prediction. So, according to the sages, the most important scientific endeavor is astrophysics, which it is a commandment to pursue.

Now, *tekufot*, which literally means "cycles," are a time image and *mazalot*, which literally mean "star constellations," or "galaxies" are a space image, which also implies that it is a *mitzvah* to unite time and space. And, once more, this appears at the end of the discussion in the Talmud regarding the tearing of the fabric of the tapestry of the Tabernacle, which we have explained corresponds to the fabric of space-time.

What *Rashi* says about science being "recognizable wisdom" is extraordinary. We mentioned earlier the cornerstone (or top stone) of the Temple through which all personal Divine Providence is focused. We noted that Zachariah describes this stone as having twice חֵן (i.e., grace, or symmetry). But, the two letter word חֵן is also an acronym for "concealed wisdom" (חָכְמָה נִסְתֶּרֶת), one of the names by which Kabbalah is referred to. So now we can add that חֵן is also an acronym for this new connotation for science that we just learnt from Rashi, "recognizable wisdom" (חָכְמָה נִכֶּרֶת). So now, we can say that these two types of wisdom are the two types of חֵן that the cornerstone (top stone) possesses.

This passage from the Talmud even says that whoever is qualified and capable and has the mind to study science according to the Torah but does not do this and only studies Torah, even if he becomes a great Torah sage, you are not allowed to say a teaching in his name. It is forbidden to learn Torah from him. One of the commentaries appearing in the *Shulchan Aruch* itself is the *Eliyah Rabbah*.[150] He tries to explain this harsh statement. He says that a person who has the ability to be wise in recognizable wisdom, in science, but does not do so, he himself has proven that he is not worthy of the title of a "wise man." So he himself has prevented himself from taking on the title of a sage. And, in principle, anyone who does not have the title of a sage you cannot repeat his teachings in his name, even if they are Torah teachings.

So we recommend strongly that this passage of the Talmud be studied in depth by everyone. God should grant us the talent and ability to learn, understand, and teach both types of wisdom, both the hidden wisdom, which is Torah, and the recognizable wisdom, which is science.

Endnotes

Lecture 1

1. To a certain extent, we are already used to science producing counter-intuitive results. Take for example the counter-intuitive idea that the earth is round and not flat. Or, that regardless of their weight, the terminal velocity of all objects as they fall towards earth is the same. Nonetheless, the physics of the 20th century has been challenging our most basic common-sense attitudes toward space, time, energy, and matter.
2. Every Hebrew word is rich in meaning. In the Bible, the Hebrew word for "world" (עוֹלָם) means "forever." Only at the very end of the Biblical period did it begin to take on the meaning of "the universe" or "the world." In the language of the sages, the same word takes on the meaning of "mind-space" or "mind-set," which are very modern words. For instance, the sages write that a person sees seven "worlds" in his life. As a newborn he is treated like royalty; as a two-year-old he begins to resemble a pig (because he's always on the ground in the dirt); at ten-years-old he jumps like a kid-goat and so on. Or, they say that King David saw five worlds in his life. He saw the world of his mother's womb, etc.
3. We have stated that counter-intuition is an example of the World of Emanation illuminating the common-sense intuition on one of the lower three Worlds. Indeed, the father principle corresponds to the World of Emanation, whereas the three lower Worlds, whose common-sense intuition can be countered, originate from the equivalent of the mother principle.
4. Work on the Babylonian Talmud officially ended in the 6th century CE, while the Jerusalem Talmud was finished about a century earlier.

5. In Hebrew, "Eve" is written חַוָּה.

   The number 19 is part of what in Kabbalah is called the Eve series of numbers. The function describing the nth Eve number is: $f[n] = 2\triangle n - 1$, where $\triangle n$ stands for the sum of integers from 1 to n. The first few Eve numbers are therefore: 1, 5, 11, 19, 29,...

6. This is attested to by the popularity of books like "Men are from Mars..." that dwell on this difference between the male and female modes of common-sense.

7. For a more complete discussion of the technical aspects of *gematria* see *What You Need to Know About Kabbalah*, pp. 72ff. and our website: www.inner.org/gematria.

8. The scientific revolution of the 17th century was made possible in part by the willingness of reformers like Galileo to give credence to the measurable quantities of physical objects and events (e.g., velocity, weight, dimensions, etc.). Until then, the quantitative attributes of a physical object were thought to be accidental, thus revealing very little insight into either its nature or expected behavior. Language too, especially the language of the Torah or the language of the sages, can be analyzed quantitatively. When this analysis is carried out correctly, it too yields insight into the nature and function of words.

9. One of the main teachings of Rabbi Elimelech of Lizhensk, a master of the third generation of Chassidut.

10. See *Tanya*, ch. 31.

11. But, just as in Escher's famous counter-spatial sketches, if you feel that you are moving up, you are really already on the way down! Instead, humility before God and feeling no better than others are the key to keeping a forward momentum going.

12. *Sanhedrin* 97b.

13. *Rosh Hashanah* 10b-11a.

14. *Avot* 3:13.

15. The three stages of every spiritual process, which are derived from an analysis of the Biblical word *chashmal* (see Ezekiel 1:27), are:

submission, separation, and sweetening. See in length in *Transforming Darkness into Light: Kabbalah and Psychology*.

16. Bertrand Russell once quipped that when a reporter asked him to explain relativity, because only three people in the world really understand it, he replied: "I am aware of Einstein and myself. Who is the third?" The reader is referred to the first chapter of Russell's *The ABC of Relativity* from which we will quote a few lines:

    > Everybody knows that Einstein did something astonishing, but very few people know exactly what it was....It is true that there are innumerable popular accounts of the theory of relativity, but they generally cease to be intelligible just at the point where they begin to say something important. The authors are hardly to blame for this. Many of the new ideas can be expressed in nonmathematical language, but they are not less difficult on that account. What is demanded is a change in our imaginative picture of the world—a picture which has been handed from remote... ancestors and has been learned by each one of us in early childhood. A change in our imagination is always difficult, especially when we are no longer young. Einstein's ideas... will seem easier to generations which grow up with them, but for us a certain effort of imaginative reconstruction is unavoidable.

17. *Sanhedrin* 24a.
18. *Ibid.* 99a.
19. *Or Torah, Shir Hashirim* 192.
20. *Megillah* 16b.
21. The inner dimension of a "vessel" is a container to absorb spiritual energy/light; the outer dimension is a "tool" for acting in reality.
22. See Zachariah 3:7.

Lecture 2

23. In modern string theory there is an equivalent concept called a zero-brane. In string theory, a zero-brane (D_0-brane) is a dimensionless point-charge that serves to connect string-charges in any spacetime dimension.
24. Quoted many times in halachic and Chassidic literature. See *Akeidat Yitzchak* 14.
25. Deuteronomy 29:28.
26. The Greek hyle, known as חֹמֶר הִיּוּלִי in Torah.
27. Einstein did not know about this distinction.
28. *Keter Shem Tov*, Supplements 227. Though here, and elsewhere, we translate the Hebrew word *etzem* (עֶצֶם) as "essence," there is no exact English translation. What *etzem* refers to is more accurately the "essential being," which always remains unchanged as it underlies all appearances and revealed characteristics. In a philosophical sense, *etzem* might best be equated with the Greek *hypokeimenon*.
29. For more on this see the series of (Hebrew) lectures on the topic of "Time and Space in the Dimensions of the Soul," given by Rav Ginsburgh. See also Eliezer Zeiger's "Time, Space, and Consciousness" in the *B'Or Hatorah Journal*, vol. 15, which is based on some of the insights offered by Rav Ginsburgh in that lecture series.

30. The four letters of God's essential Name, also known as the Tetragrammaton are: *yud* (י), *hei* (ה), *vav* (ו), *hei* (ה). For a more complete discussion see *What You Need to Know About Kabbalah*.
31. *Tamid* 32a.
32. Job 28:12.
33. See *Zohar* II, 56a; *Zohar* III, 4a.
34. More explicitly, Kabbalah teaches that the lower unification excites the higher unification to provide it with new life-force. This is similar to

the manner in which energy/matter affects space-time, which in turn affects energy/matter.

35. The *gematria* of these four words in Hebrew is 559, half the value of our declaration of faith in one God, "Hear O' Israel *Havayah* is our God *Havayah* is one" (1118 – the lowest common multiple of 26, *Havayah*, and 86, *Elokim*, the two Names of God, representing God's transcendence and His imminence, that are absolutely unified in this verse). As every particle of matter has an anti-particle, so it is with regard to energy, and today scientists speak of anti-space and anti-time as well. Taking this dual aspect of all reality into consideration doubles the value of the four basic components of reality, bringing the total value to 1118, the *Shema* – stating that all is one (for God is one and "God is all and all is God").

36. הָאוֹר דָּבוּק בְּמָאוֹר. Technically, this term is used in conjunction with the Arizal's notion of the contraction. In Chassidut, which argues that the contraction is only figurative, not literal, it is explained that though God reveals Himself in only a limited manner in the world, dubbed His light, nonetheless, the revelation, the light, is still "clinging," i.e., inherently related to the source, i.e., God Himself. Thus, God's Presence is permanent and only concealed. For a more in depth explanation of this topic, see Rabbi Ginsburgh's Hebrew volume: *Chasdei David*, vols. 11-12, *Sod Hatzimtzum*.

37. More exactly, there can be shown to be at least one theorem which is known to be derivable from the axioms, which nonetheless, because of its paradoxical nature, cannot be proven to be true based on the same axioms.

Lecture 3

38. Genesis 6:8.

39. *Pesachim* 50a, and elsewhere.

40. These colors have nothing to do with the physical colors we see with our eyes. They are simply a nomenclature used by physicists to

describe a particular property of some elementary particles, like gluons (and quarks).

41. Proposed in the 1950s by Yang and Lee and later verified by Wu. This is called the violation of parity conservation.
42. *Zohar* III, 236a. The original phrase is "two halves of a [single] body" (תְּרֵין פַּלְגֵי גוּפָא).
43. In the Arizal's writings, it is explained that in the World of Chaos victory and thanksgiving were a single *sefirah*, while in our reality, the World of Rectification, they are two separate *sefirot*. See *Etz Chayim* 8:4, compare to 9:2.
44. *Ibid.* 19:9b. This statement is usually quoted in the name of the *Zohar*. See *Tikunei Zohar* 13.
45. Exodus 38:8.
46. In the idiom, "He is in victory, she is in thanksgiving," the two *sefirot* of victory and thanksgiving are actually the origins of the male and female parts of the intellect of the *nukva*, the feminine, meaning that the feminine wisdom derives from victory and the feminine understanding derives from thanksgiving. These two *sefirot* together serve to construct the *partzuf* of the feminine figure, which is the *nukva* of *ze'er anpin*, or *malchut* (kingdom).
47. *Rashi* to Exodus 38:8.
48. See Numbers 5:11-31. Contrary to some interpretations, the main reason for using the technique described therein is to make peace between a husband and his wife.
49. See *What You Need to Know About Kabbalah*, pp. 152-3.
50. See *Nitei Gavri'el – Hilchot Nissu'in* Part 1, p. 129, notes 21 and 22.
51. *Eiruvin* 13b.
52. See *Zohar* I, 87a and *Mikdash Melech* there.
53. As it contains 42 words.
54. As in the previous case, this too is based on the number of words in the second paragraph. However, there are more than 72 words in the

second paragraph. Explains the Arizal, that the 72nd word וְשַׂמְתֶּם, is the last word that indicates a state of might.

55. As explained in Chassidut regarding the verse, "And he [Moshe] sees the vision of God" (Numbers 12:8). Moshe sees the same vision of reality that God sees, meaning he sees things from God's perspective.
56. When permuted, these four letters spell the word "Tanya" (תַּנְיָא). This was one of the reasons that the Alter Rebbe chose this word to begin his classic work of Chassidut. The word אֵיתָן, *eitan* also alludes to the essential strength of character of the Jewish soul (for which reason, Abraham was called "*Eitan Ha'ezrachi*," see I Kings 5:11 and Psalms 89:1; see in length in *The Art of Education*, pp. 70-72).
57. Deuteronomy 6:4.
58. *Shulchan Aruch, Orach Chayim* 61:8.
59. Genesis 28:14.
60. Proverbs 5:16
61. This seems to be true in English as well, where "aft," the source of "after" (indicating the future), also means the back side, while "fore," the source of "before" (indicating the past), also means the front side.
62. As the *Zohar* says, "He [God] looked in the Torah and created the world" (*Zohar* I, 134a; *Zohar* II, 161a)
63. Zachariah, ch. 5.
64. *Sefer Yetzirah* 2:4.

Lecture 4

65. See Rambam, *Sefer Hamitzvot*, 2, where this verse is defined as the Torah source for the commandment of unifying God. See also in length in *Derech Mitzvotecha, mitzvat ha'amanat Elokut*.
66. Isaiah 25:8. Death is one of the seven names given to Esau's angel, see *Sukah* 52a.

Lecture 5

67. *Zohar* III, 168b.
68. *Zohar I,* 192b etc.
69. Science has not yet proven that super-partners exist. Using our analogy of male and female, the super-partner is like the marriage partner. Every person who is still single has to believe that they have a partner somewhere out there, and that they have simply not found them yet.
70. *Kidushin* 41a. הָאִישׁ מְקַדֵּשׁ בּוֹ וּבִשְׁלוּחוֹ, the *vav* here literally means "and."
71. Song of Songs 2:5.
72. In the story told by the fifth beggar, the hunch-back.
73. Isaiah 26:9.
74. For clarification: in the Ba'al Shem Tov's teachings this is known as "the characteristic of equality" (מִדַּת הַהִשְׁתַּוּוּת). A person should relate to all of his circumstances with equanimity. Everything is equally good and meaningful, for everything that makes up one's life is directly from the Almighty. For example, if one is being mocked by other people or one is being praised by others, it does not matter. One should thank God for both and see the manner in which both are necessary and good. See *Tzava'at Haribash*, 2.

Lecture 6

75. *Avot* 6:4. See also *Rashi* to *Gitin* 17a.
76. Psalms 104:24.
77. Psalms 111:10.
78. Job 28:23.
79. Proverbs 7:4.
80. Job 28:23.

81. Psalms 82:6.
82. Isaiah 32:8.
83. Leviticus 26:13.
84. The equivalence to a triangular number is true of the letters of this word (וְהוּא), in which all the letters precede the letter *yud* in the Hebrew alphabet. But in general, *mispar keedmee* is the sum of all letter values up to and including the value of that letter.

Lecture 7

85. All particles have relative mass because in relativity everything has energy which can be translated as mass.
86. See Zachariah ch. 6.
87. We read Isaiah's vision as the *haftarah* (the portion in the Prophets) of *parashat Yitro* in which the giving of the Torah at Mt. Sinai is described. There is always some resemblance between the *parashah*, the Torah portion and the *haftarah* that the sages linked with it. In this case, the connection teaches us that the vision of the chariot seen by Moses and the entire Jewish people at Mt. Sinai (which is not explicit in the Written Torah), was the one corresponding to the World of Emanation. This is referred to as the Chariot of Moses.
88. In *Hayom Yom* (for 29th *Adar Sheini*), the Lubavitcher Rebbe mentions that the world was created as something from nothing, and the purpose of a Jew is to reveal the original nothingness within every something.
89. So our Passover cleaning is really to clean out all the negative virtual particles and to reveal the good virtual particles through the burning of the Paschal lamb.
90. Deuteronomy 8:3.
91. Or, שִׁתּוּף פְּעֻלָּה.
92. The Arizal taught that the *dalet* (ד) of "one" (אֶחָד), which in the Torah scroll is written extra large, should actually be the size of 4 letters

dalet. So this would be like 4 (ד = 4) to the fourth power. $4^4 = 16^2 = 256$, the value of Aharon (אַהֲרֹן). Indeed, Aharon the High Priest was unique amongst the priests in that he wore 8 garments of priesthood. They themselves were divided into the four simple white garments of all priests and the four golden garments reserved for the high priest.

93. Ezekiel saw the vision of the chariot that exhibits the most inter-inclusion, and therefore in a sense inter-includes the other two visions.
94. In the *Guide to the Perplexed*, Maimonides bases many of the topics on the number 4, even though he was against the use of numbers in a "magical" or "mystical" way. As can be seen throughout his writings, Maimonides' favorite number is 4. Nonetheless, Maimonides never explicitly mentions that the underlying uniqueness of the number 4 goes back to the Tetragrammaton, the four letter essential Name of God. The sages refer to this Name as the "Name of four [letters]."
95. *Pesachim* 6b.
96. Genesis 2:7.

Lecture 8

97. In Kabbalisitic terminology this is known as "the source of the vessels is higher than the source of the lights."
98. This word originally comes from the ancient Greek *hyle*, however its meaning in Jewish philosophy is markedly different. One of the most important early discussions in Jewish philosophy of the *hiyuli* matter appears in Nachmanides' commentary to the second verse in Genesis. Interestingly, today Hyle is the name of the International Journal for the Philosophy of Chemistry.
99. It may be noted in passing that the word "matter" in English stems from the root form of "mother," indicating that it refers to the revealed aspect of matter, which is feminine (referred to in Kabbalah as the lower mother).
100. Proverbs 31:30.
101. *Be'er Mayim Chaim*, *Terumah* ch. 25.

102. Zachariah 3:9.

103. Ibid. 4:7.

104. Since the Ashkenazi pronunciation does not differentiate between *alef* and *ayin*, "Ein[stein]" can also be understood as the Hebrew for "eye" (עַיִן).

105. This concept is discussed in many places in Kabbalah. The consummate work on this subject remains to date Rav Natan Shapira's *Toov Ha'aretz*. For a pdf version of this work, see: http://www.hebrewbooks.org/5921.

106. Genesis 1:4.

107. Job 38:28.

108. This is one of the principles of Hebrew grammar appearing in the Malbim's *Ayelet Hashachar*, his introduction to Leviticus.

109. Isaiah 26:19.

110. *Ketubot* 111b.

111. On the other hand, in principle every word written with an *ayin* has as its inner essence the same word written with an *alef*.

112. *Ofan* (אוֹפַן) is another type of angel seen by Ezekiel in his vision of the Divine Chariot. It literally means a "wheel," and its numerical value is 137, the approximate inverse of the fine-structure constant, a quantity inherently linked with the notion of an electron. For more on the reasons for identifying the electron with this word, see our upcoming volume on the significance of the number 137 in Torah.

Lecture 9

113. Genesis 21:8.

114. Genesis 1:4.

115. Genesis 24:1-67. From the first mention of camels in verse 10, to the final one in verse 64, there are altogether 58 verses.

116. *Avot* 5:7.

117. As explained earlier, the triangle of n is the sum of integers from 1 to n. In this case, the sum of integers from 1 to 73 is 2701. In function form $\triangle n = \frac{n \perp 1}{2}$.

118. Physicists today are fond of likening the abundance of elementary particles discovered to a particle zoo.

119. *Berachot* 56b.

120. Psalms 92:6.

121. Jeremiah 51:56.

122. *Berachot* 33a.

123. *Berachot* 54b.

124. As the sages say that one should make signs to remember one's learning (סִימָנִים עֲשֵׂה); see, *Shabbat* 104a. See also, *Yalkut Shimoni, Vayeishev*.

125. Genesis 1:20.

126. Joshua 2:11.

127. Deuteronomy 4:39.

128. *Midrash Devarim Rabbah* 2:27.

Lecture 10

129. Proverbs 2:19.

130. Psalms 119:176.

131. *Midrash Vayikra Rabbah* 29:11.

132. Psalms 88:13.

133. For example: the numerical value of "Mashiach" (מָשִׁיחַ), 358 is the same as that of "serpent" (נָחָשׁ).

134. *Shabbat* 77b.

135. Ecclesiastes 2:13.

136. Genesis 1:2.

137. *Mishnah Avot* 3:10. From the phrase, "Be very careful and guard yourself, lest you forget those things that your eyes saw" (Deuteronomy 4:9) the sages learn that a person must not forget the Torah that he has studied. However, from the second phrase in the same verse, "lest they shall be discarded from your heart, all the days of your life," they learn that the prohibition of forgetting is only if a person does so on purpose—choosing to forget his studies, implying, that if it happens naturally (תָּקְפָה עָלָיו מִשְׁנָתוֹ), it is not lost altogether and can still be retrieved.

138. Isaiah 25:8.

139. Job 20:15.

140. *Mishnah Avot* 3:15.

141. See note 124 above.

142. *Zohar* II, 7b.

143. *Sefer Yetzirah* 5:2.

144. One of the things that these 12 diagonal lines correspond to is the 12 tribes of Israel.

Lecture 11

145. *Shabbat* 74b-75a.

146. Psalms 22:7.

147. *Smag (Sefer Mitzvoth Gadol)*, positive commandment 47.

148. Deuteronomy 4:6.

149. See also the Gaon of Vilna's commentary to *Tikunei Zohar* 21.

150. *Orach Chaim* 340:14.

Index

More Books by Harav Yitzchak Ginsburgh

The Hebrew Letters
Channels of Creative Consciousness
502 pages

Transforming Darkness into Light
Kabbalah and Psychology
192 pages

Living in Divine Space
Kabbalah and Meditation
288 pages

Anatomy of the Soul
144 pages

Awakening the Spark Within
Five Dynamics of Leadership that can Change the World
200 pages

Rectifying the State of Israel
230 pages

Kabbalah and Meditation for the Nations
216 pages

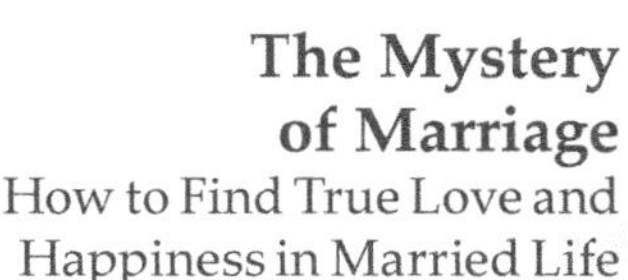

The Mystery of Marriage
How to Find True Love and Happiness in Married Life
500 pages

Body, Mind and Soul
Kabbalah on Human Physiology, Disease and Healing
342 pages

The Art of Education
Internalizing Ever-New Horizons
302 pages

What You Need to Know About Kabbalah
190 pages

A Sense of the Supernatural
Interpretation of Dreams and Paranormal Experiences
208 pages

Consciousness and Choice
Finding Your Soulmate
284 pages